Sukhoi Su-30 MKK/MK2/M2

Russo-Kitayshiy Striker from Amur

HUGH HARKINS

Copyright © 2015 Hugh Harkins

All rights reserved.

ISBN: 1-903630-18-5
ISBN-13: 978-1-903630-18-1

Sukhoi Su-30 MKK/MK2/M2

Russo-Kitayshiy Striker from Amur

© Hugh Harkins 2015

Published by Centurion Publishing
United Kingdom

ISBN 10: 1-903630-18-5
ISBN 13: 978-1-903630-18-1

This volume first published in 2015
The Author is identified as the copyright holder of this work under sections 77 and 78 of the Copyright Designs and Patents Act 1988

Cover design © Centurion Publishing & Createspace

Page layout, concept and design © Centurion Publishing

All rights reserved. No part of this publication may be reproduced, stored in a retrieval system, transmitted in any form, or by any means, electronic, mechanical or photocopied, recorded or otherwise, without the written permission of the Publisher

The Publisher and Author would like to thank all organisations and services for their assistance and contributions in the preparation of this volume. JSC Sukhoi Design Bureau (Sukhoi Aviation Holding Company), United Aircraft Corporation, NPO Saturn, United Engine Corporation, JSC V. Tikhomirov NIIP, Zhukovsky, JSC Tactical Missiles Corporation, JSC Uspensky Avionica Moscow Research and Production Complex, JSC Concern Radio-Electronic Technologies (KRET), RAC-MiG, Rostec Corporation, Sukhoi KnAAPO, Sukhoi KNAAZ and NPP Zvezda

CONTENTS

	INTRODUCTION	i
1	T-10	1
2	FROM INTERCEPTOR TO MULTI-ROLE – Su-30M TO Su-30MK	11
3	Su-30MKK – KITAYSHIY STRIKER	19
4	IMPROVING THE BREED – Su-30MK2/M2	51
5	OFFENSIVE AND DEFENSIVE WEAPON SYSTEMS	75
6	Su-35UB – SUKHOI ENIGMA	95
7	APPENDICES	97
8	GLOSSARY	100

INTRODUCTION

The Su-30MKK/MK2/M2 family of combat aircraft share the same numerical designation as the Su-30MK/MKI/SM family of multi-role combat aircraft which were clearly derived from the Su-30M, which was, itself, derived from the Su-27UB two-seat operational conversion variant of the Su-27S single-seat air superiority fighter. However, the numerical designation and some first glance looks aside, the Su-30MKK family is fundamentally a different design from the Su-30MKI family in terms of primary design role, internal systems and structure; certainly in regards to the latter the aircraft having more in common with the Su-27M (first generation Su-35) than the Su-27UB or Su-30M.

This volume covers the evolution of the Su-30 design from the Su-27, which was born out of the T-10 development program of the 1970's. The evolution of the Su-30M interceptor into the multi-role Su-30MK is described. The Su-30MKK is covered in detail, as are the improved Su-30MK2 for the export market and the Su-30M2 strike fighter developed for the Russian Federation Air Force. The aircraft and systems are described in detail, as are the weapons that can be employed by the respective variants. The enigma that is the Su-35UB is included as that aircraft, in regards to structure, has more in common with the Su-30MKK family than any other Su-27 family variant.

All technical information regarding the aircraft, systems and weapons have been provided by the respective manufacturers, as have many of the photographs and diagrams. Elements of the first and fifth chapters borrow certain texts from the volumes 'Sukhoi Su-35S 'Flanker' E, Russia's 4++ Generation Super-Manoeuvrability Fighter' and 'Sukhoi T-50/PAK FA, Russia's 5th Generation 'Stealth' Fighter'.

1

T-10

The origins of all members of the Sukhoi Su-30MKK/MK2/M2 designs go back to the Sukhoi Su-27 program, the origins of which go back to the Sukhoi T-10, design work on which commenced under the leadership of O.S. Samoilovich in late 1969 when the Sukhoi OBK Design Bureau in the Soviet Union embarked upon studies for a new long range air superiority fighter for the Soviet IA-PVO (*Istrebitelnaya Aviatsiya Protivo-Vozdushnoy Obstrany*/Air Defence Force). Requirements for the new fighter included long-range and high manoeuvrability, combined with modern radar and weapon systems to enable the aircraft to be capable of defeating the most modern western fighter aircraft then in service or development, typified by the McDonnell Douglas (later Boeing) F-15 Eagle then under development in the United States under the FX program. A secondary ground attack role was stipulated for the new Soviet fighter design.

A number of concepts were studied, various designs being drawn up by the design houses of P.O. Sukhoi, A.I. Mikoyan and A.S. Yakovlev during 1971-1972. The two former design houses were designing high agility aircraft, both arriving at similar configurations, apparently because both utilised data from the same research agency. The design that eventually became the PFI (Advanced Frontline Fighter) Project 9, therefore, resembled a scaled down version of the Sukhoi T-10, the latter being designed around a highly blended fore-body and high lift ogive wing with LERX (Leading Edge Root Extensions).

The initial design for the T-10 was complete by September 1971, submitted in February 1972, and, following a preliminary review, design revisions were incorporated, following which full-scale development commenced in conjunction with development of a lightweight fighter by Mikoyan; the MiG-29 (Project 9). The Sukhoi and Mikoyan designs were, however, not in competition with each other; the Su-27 being planned as a heavy fighter and the MiG-29 being planned as a light fighter capable of engaging its NATO opposite numbers, later typified by the General Dynamics (later Lockheed Martin) F-16 Fighting Flacon.

Artist rendering of a Su-30MK derivative in Russian service. KnAAPO

During the course of 1972-73, the T-10 was further redesigned with increased wing area and fuel capacity. The thrust of the proposed powerplant, A.M. Lyulka (NPO Saturn) AL-31F (Article 99), was increased to compensate for increased weights. By 1975, Sukhoi data described the T-10 (Su-27) design with the following features, "an integrated ogive wing configuration, leading-edge root extensions, an all-moving horizontal tail unit mounted on the centre wing section continuation beams, and twin tail fins mounted on engine nacelles at the airframe stern post". The variable engine intakes were positioned "either side of the plane's roll axis, and suspended from the centre wing section" ensuring a stable air flow to the engines even when the aircraft was flying at high AoA (Angle of Attack).

The T-10 was designed with inherent lateral instability balanced by an EDCS (Electronic Distance Control System), computerised FBW (Fly-By-Wire) FCS (Flight Control System). The Soviet Union had pioneered FBW technology with the Sukhoi T-4 intermediate range bomber prototype (cancelled in the 1970's), which conducted its first flight as Black 101 on 22 August 1972. The first American combat aircraft designed with a FBW FCS, the General Dynamics YF-16, conducted its maiden flight on 2 February 1974, followed by the first flight of the European Panavia Tornado prototype (then known as the MRCA - Multi-role Combat Aircraft) on 8 August that year. While the T-4 flew before the YF-16 and Tornado, the FBW FCS installed in the Western aircraft were more advanced than the Soviet system, providing FBW control in all axes.

The prototype Sukhoi T-10, T-10-1. Sukhoi

Approval of the T-10 (Su-27) configuration was granted on 19 January 1976. Three prototypes (two flight and one ground test) were under construction by early 1976, the first, T-10-1, being completed in April 1977. This aircraft, along with the second, third and fourth development aircraft, T-10-2, T-10-3 and T-10-4, was built at Sukhoi's experimental plant near Moscow.

Western intelligence agencies got their first glimpse of the new Soviet fighter design from photographs taken by a spy satellite while the aircraft was on the ground at Zhukovsky (then known in the west as Ramenskoye). In normal NATO (North Atlantic Treaty Organisation) practise the aircraft was allocated the reporting name RAM-K, as it was the tenth experimental fighter observed at the base (the letter I was apparently not used in the RAM reporting sequence). The MiG-29 (Project 9) was allocated the reporting name RAM-L.

The prototype T-10, which was powered by a pair of AL-21FZAI afterburning turbojet engines, flew for the first time on 20 May 1977, piloted by Sukhoi Chief Test Pilot V.S. Ilyushin. The AL-21FZAI, which was an interim engine rated at 76.49 kN (16,195 lb.) dry and 109.84 kN (21,692 lb.) in afterburner (the available thrust may have been a little higher than these interim ratings), was a derivative of the AL-21F-3 engine that was used to power a number of Soviet combat aircraft including the Su-17 'Fitter' and the Su-24 'Fencer' variable-geometry strike aircraft. While the dry thrust of the AL-21F-3 was higher than the dry thrust of the AL-31F (Article 99) planned for the production Su-27, the afterburner thrust was lower and the engine had a much higher specific fuel consumption compared with the AL-31F; a serious consideration for an aircraft designed as a long-range air superiority fighter.

Amongst the first images of the Sukhoi T-10 design to emerge, this grainy still of the first prototype, coded 01, allowed Western intelligence agencies a tantalising glimpse of what the new generation Soviet air superiority fighter aircraft looked like.

Following construction of the initial batch of four aircraft a further five, T-10-5, T-10-6, T-10-9, T-10-10 and T-10-11, were built at the Komsomolsk-on-Amur production plant. From the third prototype, T10-3, power switched to the more powerful (in afterburner thrust) AL-31F turbofan developed to power the production standard aircraft. Normal thrust ratings for this engine were given as 79.43 kN (17,857 lb.) dry and 122.59 kN (27,558 lb.) with afterburner. Available information indicates that the AL-31F has a nine-stage HP (High-Pressure) compressor, a four-stage LP (Low-Pressure) compressor and cooled single-stage HP and LP turbines to the rear of the combustor. The efficient air flow afforded by the combination of engine technology, the aircraft air intake design and computer controlled variable inlet guide-vanes, contributed to the Su-27 high performance, conveying varying degrees of capability to conduct extreme high alpha manoeuvres like the 'Cobra' or Tail Slide without the engines stalling.

When it entered service in the 1980's, the high thrust to weight ratio of the AL-31F bestowed upon the standard Su-27S a high maximum speed, unrivalled (for the time) supersonic acceleration, climb rate and manoeuvrability, in certain flight regimes such as sustained turn rate, for an aircraft in its class. Typical engine life was set at around 3,000 hours with a TBO (Time between Overhaul) of 1,000 hours; reasonable figures for a Soviet era fighter engine. It should be noted that AL-31F engines have been run for thousands of hours over their scheduled life expectancy during bench running tests.

A number of problems with performance goals were encountered during the T-10 flight test program resulting in the decision to implement a more or less complete redesign of the aircraft to address issues such as controlling weight, reducing drag, increase wing lift and improving roll control. The T-10-7 was therefore built as the prototype of the new design, receiving the new designation T-10S-1, which in turn received the NATO reporting name 'Flanker' B; the original T-10 aircraft having been allocated the NATO reporting name 'Flanker' A. The T-10S-1, which was more or less a new design, was flown for the first time on 20 April 1981.

When NATO aircraft began encountering the Su-27 from the mid-1980's it was noted that the aircraft carried new generation air to air missiles which emerged as the R-27 (K-27). The R-27 was produced in SARH (Semi-Active Radar Homing), infrared homing and radiation (radar energy) homing variants; the SARH variant was then thought to have a similar capability to late production AIM-7 Sparrow SARH air to air missiles then arming the Su-27's USAF counterpart - the F-15 Eagle. US DoD

The T-10S-1 had a new tapered wing with a straight, slatted leading edge flap, flaperon and cropped wingtips incorporating missile launch rails that doubled as anti-flutter weights. The flaperons and differential tailerons replaced the ailerons of the original T-10 design. It was the changes to the fuselage that were most noticeable, with a shallower, longer drooping nose and deeper spine. The twin vertical tail fin configuration of the T-10 was retained, but this was moved outboard from their original position on top of the engine nacelles to booms, which lay alongside the engines. The main undercarriage door mounted air brakes of the T-10 were replaced by a single spine mounted unit similar to that seen on the F-15 Eagle. The new main undercarriage was repositioned, as was the nose wheel, which was moved aft.

When the T-10S-1, formerly the incomplete T-10-7, conducted its maiden flight on 20 April 1981 (pilot V.S. Ilyushin), it was clear that it was a new design. However, even with the new, design, problems were encountered during flight testing, especially with the wing. The solution to this problem was to reduce the area of the leading-edge slats.

The Su-27 'Flanker' B remains in widespread Russian Federation Air Force service in the second decade of the 21st century. This aircraft was encountered by RAF Eurofighter Typhoons over the Baltic Sea on 17 June 2014. Crown Copyright

The new design that evolved into the production Su-27S was equipped with a modern weapons system based around the RLPK-27 weapon control system, featuring a powerful N001 pulse-Doppler radar that was allocated the NATO reporting name 'Slot Back'. This system had a reported detection range of around 240 km, although manufacturer's information suggests 150 km against a fighter size target. The radar can track ten targets simultaneously, but once locked onto a target it cannot continue to scan for others. The radar was complemented by an electro-optical complex consisting of an OEPS-27 Electro-Optical Sighting System; an OLS-27 Optical Location System - IRST (Infra-Red Search and Track) and LR (Laser Range-finder), allowing the detection, tracking and engagement of targets passively without the need for radar, the emissions of which can betray the host aircraft position. A Shchel helmet mounted target designation system allowed engagements of off-boresight targets up to 60° by cueing sensors; the missile tracker head onto targets that had not been bore-sighted.

Once the design of the new fighter was finalised the Su-27S entered production (the first series production Su-27S conducted its maiden flight on 1 June 1982 with Sukhoi test pilot A.N. Isakov) and entered service in June 1985, apparently with the 60th IAP-PVO (FAR). Although having been in service for over five years the Su-27 was officially signed into service by the Soviet government on 23 August 1990.

The Su-27UB featured a second cockpit to undertake the operational conversion of pilots. This example is adorned with the colourful livery of the Russian Knights aerobatic display team. Sukhoi

Su-27S series production aircraft, which were powered by a pair of AL-31F turbofan engines, each rated at 79.43 kN dry and 122.58 kN with afterburner, have a length of 21.9 m, height 5.9 m and wingspan 14.70 m. Normal take-off weight was set at 23400 kg (Su-27SK) with 2 x R-27R1, 2 x R-73 air to air missiles and 5270 kg of fuel. Maximum take-off weight is 30450 kg (Su-27SK). The Su-27 carries 5270 kg of fuel at normal load and 9400 kg at maximum fuel-load. The huge volume of fuel allowed an impressive range to be attained; the aircraft being capable of flying 1340 km at sea level armed with 2 x R-27R1 and 2 x R-73 missiles. In the same configuration, range is 3530 km at altitude. Payload, which can be carried on ten wing and fuselage stations, is 4430 kg (according to manufacturer information, although other reports suggest up to 6500 kg) which can include the primary armament of R-27 (NATO reporting name AA-10 'Alamo') semi-active radar homing and infrared homing air to air missile variants, and R-73 (NATO reporting name AA-11 'Archer') infrared guided air to air missiles, as well as unguided air to surface munitions for the secondary air to surface role.

Although a large heavy fighter, the Su-27 showed itself to have an exceptional performance, in many areas, such as range, manoeuvrability and climb rate, being superior to its rivals. The airframe has a +9 g limit that can be over-ridden by switching the limiter off. Maximum level speed is stated as 1400 km/h at sea level and Mach 2.35 at altitude; climb rate is stated as 19800 meters per minute at sea level, with an operational ceiling of 18500 m.

Following its introduction to service with the air forces of the Soviet Union in June 1985, production continued, with in excess of 500 Su-27's thought to have been produced by the time the Soviet Union began to crumble towards the end of 1991. Following the break-up of the Soviet Union the Su-27 remained in service in Russia, assuming greater importance as older aircraft were retired. It is estimated that around 300 Su-27's remain in service with Russia. Smaller numbers of Su-27s equipped the air forces of some former Soviet Republics.

The two-seat Su-27UB operational conversion trainer retained the full combat capability of the single-seat Su-27S, albeit with a slight degradation in overall flight performance. The two seat aircraft would eventually be developed into the Su-30 interceptor which would go on to form the basis of the plethora of multi-role Su-30MK variants such as the Su-30MKI. Sukhoi

Like most single-seat tactical combat aircraft of its era a two-seat operational conversion trainer variant of the Su-27 was developed to ease the conversion of pilots onto the new fighter; development of this variant commencing in 1976. Designated T-10U-1, the prototype two-seat operational conversion trainer, known in service as the Su-27UB, was based on the production version of the single seat Su-27S. The main external differences were a redesigned forward fuselage incorporating of a second cockpit that was raised above the front cockpit affording a better forward view for the occupant in the rear cockpit. A single piece canopy covers the twin-cockpit, with a one-piece windscreen ahead of the front cockpit. The vertical tail planes and air brake were of increased height and area; the Su-27UB standing 500 mm taller than the Su-27S. Compared to the Su-27S, maximum take-off weight was increased from 30450 kg to 33000 kg, which, along with the extra drag caused by incorporation of the second cockpit, resulted in the overall performance being somewhat degraded compared to the Su-27S; maximum speed being reduced from Mach 2.35 to Mach 2.0 at altitude. Range was also reduced from 1340 km to 1270 km at sea level and from 3530 km to 3000 km at altitude. There was only a slight degradation in turn rates, initial climb rates and take-off and landing performance.

The first flight of the prototype, T-10U-1, took place on 7 March 1984, and the first series production Su-27UB, which was allocated the NATO reporting name 'Flanker' C, conducted its maiden flight on 10 September 1986, with deliveries commencing in 1987.

China, which purchased a large number of the Su-27SK and Su-27UBK (export variants of the Su-27S/UB) was the first export customer. The standard Su-27SK export variant was also purchased in small numbers by Vietnam (6) and Indonesia (2). Su-27S/UB fighters were also operated by a number of former Soviet Republics and a number of surplus Russian and former CIS (Commonwealth of Independent States), former Soviet Republics, fighter aircraft were delivered to several other nations.

Other variants developed from the basic Su-27 included the naval Su-27K (officially designated Su-33 in 1998), the Su-27M (original Su-35), Su-34 side-by-side twin seat strike aircraft, the side-by-side two seat Su-27KUB which was designed as a conversion trainer for the Su-33 (Su-27K) naval fighter, the various Su-30 variants and the second generation Su-35 known as the Su-35S. Of these much of the design work for the Su-27M would later be carried over to the Su-30MKK program.

The Su-27M, which adopted a canard-triplane configuration like that of the Su-33, emerged in the 1980's, when, even as the first generation Su-27 was entering service in the Soviet Union, plans were already underway to introduce an improved, true multi-role variant of the aircraft to operational service in the 1990's. The main aim of the Su-27M was to introduce more modern avionics and fire control systems, better close in agility and greater air to surface capability than that possessed by the first generation Su-27. Increased agility and better control and handling characteristics were achieved through the introduction of canard foreplanes like those on the Su-27K, combined with a new quadruplex FBW FCS. While the first generation Su-27S used FBW control in pitch only, the Su-27M used FBW control in all axes, with four longitudinal channels in pitch and three in roll/yaw. To compensate for the effect of the canards the vertical tail area was increased on some of the Su-27M development aircraft; the new tail fins being similar to the increased height vertical tail introduced with the Su-27UB, but with the rudder extended downward to the fin base and the fin tops squared off.

By late 2009, forty eight Russian Federation Air Force Su-27's had been modernized with an avionics upgrade, receiving the designation Su-27SM. Twelve new build Su-27SM3's (above), ordered in 2009, featured a strengthened structure, allowing up to three tons higher take-off weight to be attained over that of the Su-27S(II).

The increased weight of the Su-27M was to be partially offset by increased thrust derivative AL-31F engines which featured a number of improvements, including the incorporation of a FADEC (Full Authority Digital Engine Control) system. Although featuring a number of improvements in areas of avionics, power, range, agility and weapon delivery, Russia's post-Cold War economic downturn meant no large production orders were forthcoming, the program being cancelled.

Although the Su-27M single-seat multi-role fighter fell into abeyance, followed by outright cancellation in the early 2000's following its deselection in a number of export fighter competitions, Sukhoi continued single-seat T-10 design work with a number of upgrade programs such as the Su-27SKM and Su-27SM, the latter for the domestic fleet. Lacking funds for any significant purchases of new combat aircraft during the 1990's and into the first decade of the 21st century, Russia embarked upon a number of upgrade programs for its combat aircraft fleet with several batches of Su-27's being brought to Su-27SM standard featuring a similar standard of weapons control system, including upgraded N001VEP radar, as that installed in the Su-30MK2.

As well as upgrading some existing Su-27S to SM standard, twelve new build Su-27SM3 fighters were delivered to the Russian Federation Air Force early in the second decade of the 21st century as the service prepared to take delivery of the second generation Su-35; the Su-35S 4++ Generation 'Super-Manoeuvrable' fighter, which commenced in 2014, partially modernising the Russian fighter fleet as it awaited the introduction of a fifth generation fighter aircraft in the shape of the Sukhoi T-50/PAK FA towards the end of the second decade of the 21st century.

2

FROM INTERCEPTOR TO MULTI-ROLE – Su-30M TO Su-30MK

Now evolved into a true multi-role combat aircraft, the Su-30 began life as a low cost, low risk interceptor variant of the standard Su-27UB 'Flanker' C. In the mid-1980's, just as the two-seat Su-27UB was being developed, Sukhoi embarked upon design work for a two-seat long-range interceptor variant.

One of the many Su-27 derivatives revealed during the early 1990's was the T-10PU (Su-27PU/Su-30), which, at first glance, differed little in appearance from the Su-27UB. The most noticeable differences were the provision for a retractable in-flight refuelling probe on the port side forward fuselage just forward of the windscreen, which, as with the naval Su-27K (Su-33), and the Su-27M, necessitated having the OLS-27 sensor ball offset to starboard. The Su-30 (Su-27PU) was powered by the same AL-31F turbofan engines that powered the Su-27S/UB.

The Su-30 was initially developed to meet an IA-PVO requirement for an interceptor featuring extremely long-range to enable it to undertake patrols of the Soviet Union's vast borders, as well as providing air cover for naval forces. One of the main requirements was for a 10 hour endurance, resulting in an aircraft with twin cockpits housing two pilots, either one of which could take command of the aircraft at any given stage during the mission. As existing aircraft projects were being starved of funds Sukhoi stressed the in-house designation of Su-30, indicating that it was a new program, in order to try to secure continued funding for the project. Among the systems incorporated into the design was a new navigation system developed by RPBK Ramenskoye, which combined laser-gyro inertial sensors with the GLONASS (Globanaya Navigozionnaya Sputnikovaya Sistema - Global Navigation Satellite System) and Loran.

As well as being tasked with long-range interception the Su-30 was also to be capable of acting as a quasi-mini-AWACS (Airborne Warning and Control System)/command post, operating in conjunction with up to four other interceptors such as the Su-27S or other Su-30s. In this role the Su-30 was to be capable of automatically controlling and directing the other aircraft to the most suitable targets, whilst also receiving and transferring information such as radar and other sensor inputs within the formation via a datalink. As with the MiG-31 'Foxhound' long-range interceptor that pioneered this type of operation, a fighter controller would have been carried in the rear seat.

The prototype Su-30, which was converted from a Su-27UB, was designated T-10PU-5 with the side code Blue 05. This aircraft conducted its post conversion maiden flight in 1989. A second prototype was designated T-10PU-6 and allocated the code Blue 06. Minimal exterior design work was required for the Su-30 airframe, which was more or less identical to the standard Su-27UB, the only notable difference being the retractable in-flight refueling probe located on the port side forward fuselage just ahead of and below the windscreen, which required the OLS-27 station ahead of the windscreen to be offset to starboard instead of occupying the centre position as on the Su-27UB. Sukhoi

Early operational tests showed that the aircraft had a range in excess of 3000 km, and this could be further increased with in-flight refuelling, making the aircraft ideal for operations in the vast expanse of Northern Russia where radar coverage is poor and extremely adverse weather conditions made navigation difficult. The PVO stipulated a requirement for an aircraft with a ten hour endurance and this capability was demonstrated on 6 June 1992 when a Su-30 was flown to the North Pole and back in a time of 12 hours, employing three separate in-flight refuelling top-up's.

The prototype Su-30, T-10PU-5, Blue 05, during an early test flight in the early 1990's.
Sukhoi

Like the Su-27UB 'Flanker' C, the airframe of the Su-30 was constructed primarily of strengthened aluminium alloys and titanium. However, an undisclosed percentage of the airframe was constructed using composite materials. The T-10PU prototypes featured a modified FBW (Fly-By-Wire) FCS (Flight Control System) and enhancements to the avionics and fire control system, apparently including twin target engagement capability for the N001 radar which was otherwise the same as the system installed in the Su-27S/UB. Integrated with the radar was the OLS-27 Optical Location System that combined an IRST (Infrared Search and Tracking system) with a LR (Laser Range finder). Being almost aerodynamically identical to the Su-27UB the performance of the T-10PU was more or less identical to the former. In Russian service primary armament was to be the R-27R1 semi-active radar homing, R-27T1 infrared homing and R-73E short-range infrared guided air to air missiles. Later extended range ER1 and ET1 variants of the R-27 would be incorporated.

The two prototypes, T-10PU-5 and T-10PU-6, coded Blue 05 and Blue 06 respectively, were converted from a pair of production standard Su-27UB operational trainers, with conversion being completed around summer 1988. The Sukhoi in house designation was apparently Izdeiye 10-4-PU. The first of the prototypes, T-10PU-5, was flown for the first time from Sukhoi's Irkutsk plant in

Southern Russia on 30 December 1989. Following a program of preliminary flight testing T-10PU-5 and T-10PU-6 were transferred to the LLI test centre for further testing, following which the design was cleared for production with the service designation Su-30M.

The prototypes were followed by a small batch of production aircraft, which included a pair built for the Jupiter Insurance Company, both of which were operated alongside a single Su-27P with the Test Pilots display team based at the Gromov Flight Research Centre. These two aircraft were apparently the first two series production Su-30M, despite their Su-27PU designation. The aircraft, which were coded White 596 and White 597 and lacked operational mission equipment, flew their public display debut at the MosAeroShow in August 1992.

A modest number of Su-30 interceptors were delivered to the Russian Federation Air Force in the 1990's. These aircraft were designated Su-30M, but are also sometimes referred to by the export designation of Su-30K. Sukhoi

By early 1996, around three Su-30M's were thought to have been delivered to the PVO's Savotsleyka air base near Nizhniy Novgorod; one of the Russian air forces main advanced training bases. The Irkutsk plant is thought to have assembled up to thirty Su-30M's for PVO service by 1997, and at the Mosero air show in 1999, Sukhoi exhibited a new Su-30M for the Russian Federation Air Force, indicating that production was continuing. The exact numbers actually completed for the Russian Federation Air Force is hard to accurately determine and may actually be less than the aforementioned number of thirty.

The main operator of the Su-30M was apparently the IA-PVO's 148th Operational Conversion Unit (TsBP I PLS) combat conversion training unit based at Savotsleyka air base.

Su-30MK demonstrator, code 603, positions for take-off at Farnborough International in September 1996. Author

As the small number of Su-30's began establishing themselves in the Russian Federation Air Force in the mid-to-late 1990's, a number of training exercises were conducted with other airborne assets including the large Mach 2.83 capable MiG-31 'Foxhound' long-range interceptor. In summer 1998, four Su-30M's and two MiG-31's flew missions lasting up to 10 hours, which included a number of practise intercepts in Russia's vast northern air space. Tactics employed during these joint operations, which also apparently included a number of live air to air missile launches against drone targets, required the MiG-31's to fly ahead of the Su-30's using their more powerful N007 (Zaslon) phased-array radar to detect targets. Once targets were detected, at ranges out to 200 km, the target information was passed to the Su-30M's which would be flying some 50-60 km behind the MiG-31's. These missions could cover an area of 8000 km, requiring the use of in-flight refuelling aircraft, which were typically Ilyushin Il-76 airborne tankers. The operations were also supported by Ilyushin E-50 AWACS aircraft.

With Russia's post-Soviet Union economic plight in the 1990's ruling out any large production run for domestic use the export market was seen as the best prospect for the Su-30. However, if the Su-30 was going to have any realistic chance of being a success on the export market it was clear that additional multi-role capability would have to be incorporated into the design. The planned multi-role export variant emerging as the Su-30MK, which weighed in at around 34000 kg at maximum take-off weight; 1000 kg heavier than the standard Su-27UB at maximum take-off weight, depending on equipment fit. In the Su-30MK the back seat occupant was to be a weapons system operator supervising target selection and weapon delivery as well as overseeing operation of the on-board radar.

While the Su-30MK was being developed, an interim export variant, the T-10-4PK, with the service designation Su-30K, was offered to potential customers. This was basically a standard Su-30 interceptor with slight changes to some systems, including the IFF (Identification Friend or Foe), and the communications and navigation suite, an interim specification to allow early deliveries while the more advanced multi-role Su-30MK was developed.

Su-30MK demonstrator, side code 603, in the static park at Farnborough International in September 1996. The port side retractable in-flight refuelling probe and offset to starboard OLS are clearly evident. Author

Flight testing of the baseline Su-30MK was completed in 1993 and the aircraft was offered on the export market from 1993/94, the first real interest in the aircraft being expressed by India in the latter year.

Eventually the Su-30 would become an export success with a plethora of multi-role variants commencing with the baseline Su-30K, which, as noted above, was an interim model while India awaited delivery of the more advanced Su-30MKI, which was designed as a super-manoeuvrable multi-role fighter equipped with advanced systems, canard-foreplanes and a strengthened and modified rear fuselage able to accommodate a thrust vectoring engine nozzle; the prototype of this variant conducting its maiden flight in July 1997.

The Su-30MK demonstrator, side code 603, commences its take-off run (top) and lands (bottom) at Farnborough in September 1996. Author

India received the first of eighteen Su-30K's ordered, in 1997. The first Su-30MKI was delivered in 2002, with orders apparently standing at 272 aircraft in 2015; similar aircraft have been supplied to Malaysia and Algeria, each with differing equipment fits and a domesticated variant designated Su-30SM is in service with the Russian Federation Air Force and Navy as well as Kazakhstan.

Russia embarked upon an upgrade program for its Su-30M interceptor fleet under the initial designation Su-30KN. This added an enhanced air to surface capability with advanced precision guided munitions, and the aircraft was to be capable of operating with the Vympel (JSC Tactical Missiles Corporation) RVV-AE (R-77) active radar guided beyond visual range air to air missile.

Top: An Indian Air Force Su-30K. IAF. **Above: The Su-30KN demonstrator.** Irkut Corporation

3

SU-30MKK – KITAYSHIY STRIKER

By the beginning of the 1990s, China's PLAAF (Peoples Liberation Army Air Force) large interceptor/fighter bomber force was in drastic need of modernization. The service was then reliant on large numbers of obsolete types such as the Shenyang F-6, Chengdu F-7, Shenyang F-8I and the Fantan Q-5, all derived from such archaic designs as the 1960's era Soviet MiG-21 and 1950's era MiG-19. Even the cancelled Peace Pearl program upgrade of the F-8II would have produced little more than a McDonnell Douglas (now Boeing) F-4 Phantom II, 1960/1970's generation aircraft, albeit with a 1980's era radar system in shape of the Westinghouse (now Northrop Grumman) AN/APG-66 as fitted to the F-16A/B Fighting Falcon. This program ended abruptly in 1989 when the United States withdrew following the Tiananmen Square demonstrations in China.

Unable to gain access to American technology China turned to its Northern neighbour, the Soviet Union, which was in a position to supply either the MiG-29 'Fulcrum' or the Su-27 'Flanker', the latter type considered to be most suited to the PLAAF need for a long-range air superiority fighter aircraft; the Su-27SK/UBK (the export designations for the Su-27S/UB) eventually being selected over the lighter MiG-29. China, therefore, became the first export customer for the Su-27 when, in August 1991, that nation ordered twenty four Su-27SK/UBK's; another batch of twenty two aircraft, single seat Su-27SK and two seat Su-27UBK variants, being ordered in early 1995. The deal also included licence production of further Su-27's at China's SAC (Shenyang Aircraft Corporation) from kits supplied from Russia; at least seventy six being reported built under the designation J-11 - Jianjiji-11/Fighter Project 11 - (105 kits were supplied, but is unclear if all of these were assembled). China also purchased a further batch of forty Su-27UBK two-seat aircraft, deliveries of which were completed in late 2002. The number of Su-27's actually built by China remains something of an enigma; the uncertainty being in regards to the numbers of domestic built J-11 models.

Prior to its purchase of Su-30MKK multi-role strike fighters, the PLAAF already operated twin-seat Su-27's in the shape of the Su-27UBK which were ordered in several batches, deliveries of which were completed in 2002. US DoD

The first batch of twelve aircraft, which comprised eight single-seat Su-27SK, and four Su-27UBK two-seat operational trainers, left Chita in Russia on 27 June 1992, arriving at the PLAAF 3rd Air Division base at Wuhu the same day. The remaining ten Su-27SK and two Su-27UBK's of the initial order were delivered on 8 November 1992, apparently along with a further two aircraft donated by Russia for research and trials work, bringing the number of Su-27's delivered to 26. The PLAAF apparently initially based its Su-27's on Hainan Island (unconfirmed) before moving them to a base on the mainland. Deliveries of the second batch, ordered in 1995, commenced that year and were completed in 1996. Deliveries of a further batch of 40 Su-27UBK's, which were completed in 2002, brought the number of Su-27 deliveries from Russia to seventy six.

Plans for Chinese licence production of the Su-27SK, agreed in 1991, were apparently drawn up in October 1993, when it was decided that Shenyang Aircraft Corporation would build the type as the J-11; production contracts for license built J-11's were then placed in November 1993. Later agreements between Russia and China covered the licence production of up to 200 aircraft, but delays to the program resulted in the first Chinese built J-11, assembled from parts produced at KnAAPO (Komsomolsk-on-Amur Aircraft Production Association), not taking to the air until December 1998. As noted above, only 105 J-11 kits appear to have been supplied before the program was suspended.

While the Su-27 acquisition partially modernised the PLAAF's interceptor fleet, China was watching the development of the Su-30 program, particularly that designs potential as a strike aircraft. The Su-27SK/UBK had offered the PLAAF only a limited ground attack capability with unguided munitions. As the PLAAF sought to further modernise its combat aircraft fleet it expressed a formal interest in the multi-role Su-30MK, and in 1997, China expressed interest in purchasing a batch of Su-30 aircraft, which would be adapted to the PLAAF's strike aircraft requirement.

In the early 2000's, the most visible example of the Su-30MKK was White 502, the second of the four Su-30MKK prototype development aircraft. In addition to flight test duties this aircraft was utilised by Sukhoi for appearances at international trade shows. Author

Development of the Su-30MKK (the second K in the Su-30MKK designation stands for *Kitayshiy*, which means Chinese in Russian, although as with the Indian Su-30MKI, Sukhoi refers to the aircraft simply as Su-30MK) was launched by KnAAPO in 1997 under the design leadership of A.I Knyshev. Whereas previous models of the Su-30 had been produced at IAPA (now Irkut Corporation), the Su-30MKK, being a very different aircraft structurally, was designed by, and built at KnAAPO, which had been the main production centre for single-seat Su-27 variants, the plant also apparently having previously built a few two-seat Su-27UB's, although the bulk of this latter variant was produced at Irkut.

KnAAPO completed the basic design during 1997-1998, following which construction of a prototype commenced. This new model was fundamentally different from previous Su-30 variants, it being no exaggeration to state that the Su-30MKK was derived not so much from the Su-30M (Su-30K), which was itself derived from the Su-27UB, but more from the Su-27M (original Su-35), much of the structural design of that aircraft being carried over to the Su-30MKK. The Su-27M wing centre section, wing panels, air intakes, tail beams, vertical tail fins and undercarriage were all incorporated into the Su-30MKK without any redesign work. The Su-30MKK also incorporated the tail end fuselage assemblies of the Su-27SK, further reducing design and development work and ultimately development costs; the only significant area of the aircraft requiring newly designed components being the nose section. The design utilised a modified variant of the FCS (Flight Control System) installed in the Su-27UBK.

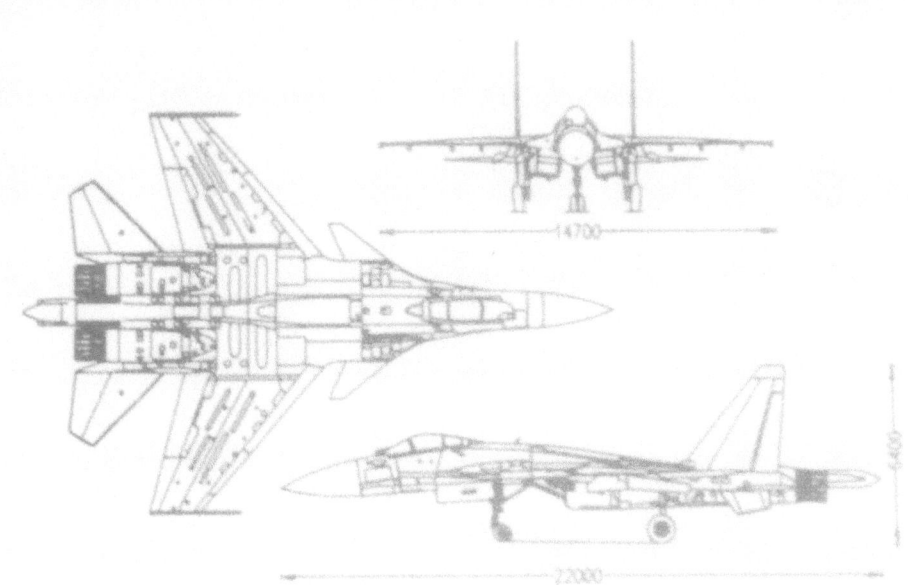

In order to fast track the Su-30MKK program, as well as significantly reducing development costs, KnAAPO drew on much of the design work for the Su-27M (original canard-tri-plane Su-35) single-seat 4+ generation multi-role fighter shown above. Author

A strengthened structure allowed the aircraft to take increased weights up to 34500 kg (34.5 metric tons) at maximum take-off weight (Sukhoi has also issued absolute maximum figures of 38800 kg). Normal take-off weight was set at 24900 kg, although this could obviously vary to a certain degree depending on the equipment fit of the aircraft. These increased operating weights allowed the Su-30MKK to conduct operations with a maximum external stores load of 8000 kg whilst carrying a full internal fuel load in line with China's requirement for such a capability.

As well as being structurally different from the Su-30MKI family, the Su-30 lacked the formers canard foreplanes or thrust-vectoring nozzles for the AL-31F engines. The lack of such canards or vectoring nozzles, along with the taller vertical tail fins, which were constructed using carbon fibre reinforced plastics, make the Su-30MKK easy to differentiate from the Su-30MKI.

Among the most visible design traits carried over from the Su-27M (original Su-35) program was the vertical tail fins and undercarriage with twin nose-wheels seen on Su-27M (Su-37MR) 711 top and above respectively. Author

Top: The Su-30MKK features design lineage from a number of Sukhoi Su-27 family variants; notably the Su-27M (original Su-35) that was developed in the late 1980's and into the 1990's, the Su-27SK (export variant of the Su-27S) and of course the Su-30M/MK. The result was in effect a hybrid Su-30/Su-27M/Su-27SK design, albeit without the latter's canard-triplane layout and equipment fit. Sukhoi

Above: This port side-on view of White 502 shows to advantage the taller squared off vertical tail fins that came to characterise the Su-30MKK/MK2 family. Author

Previous page top: A Model of the Su-30MKK exhibited at Le Bourget, Paris, in 2001. Previous page bottom: Su-30MKK vertical tail fins with wing leading edge moving surfaces in the foreground. This page: The Su-30MKK engine intake design was carried over from the Su-27M without any redesign. Author

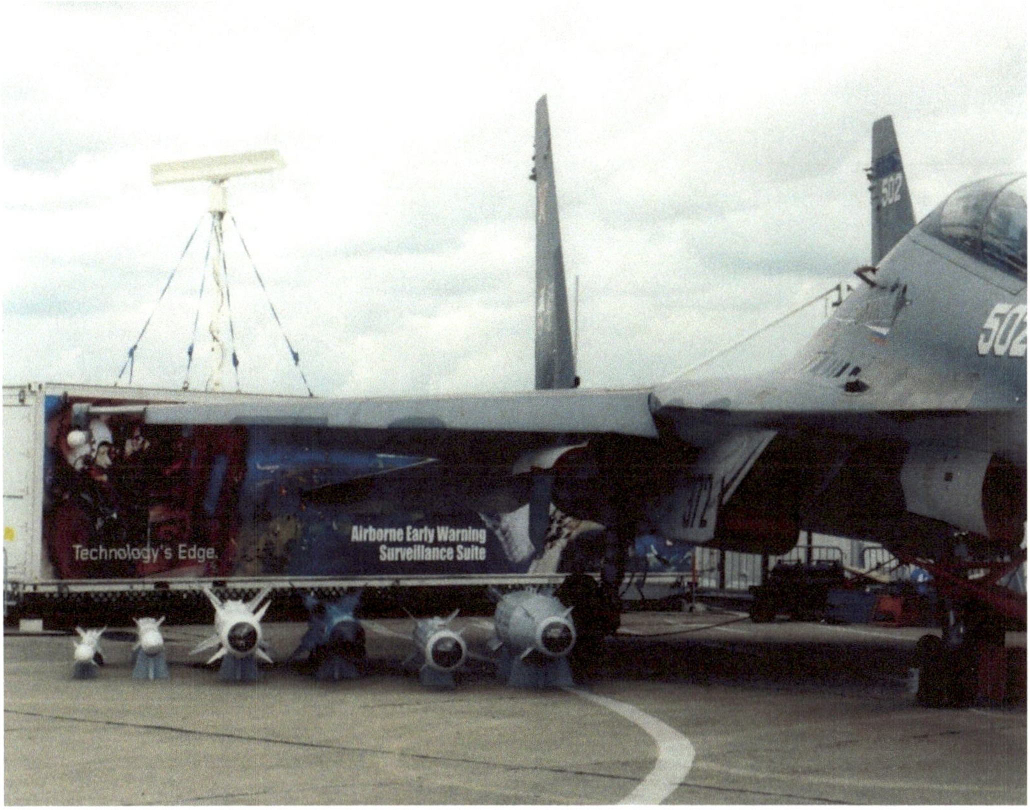
Technology's Edge.
Airborne Early Warning
Surveillance Suite
502

502

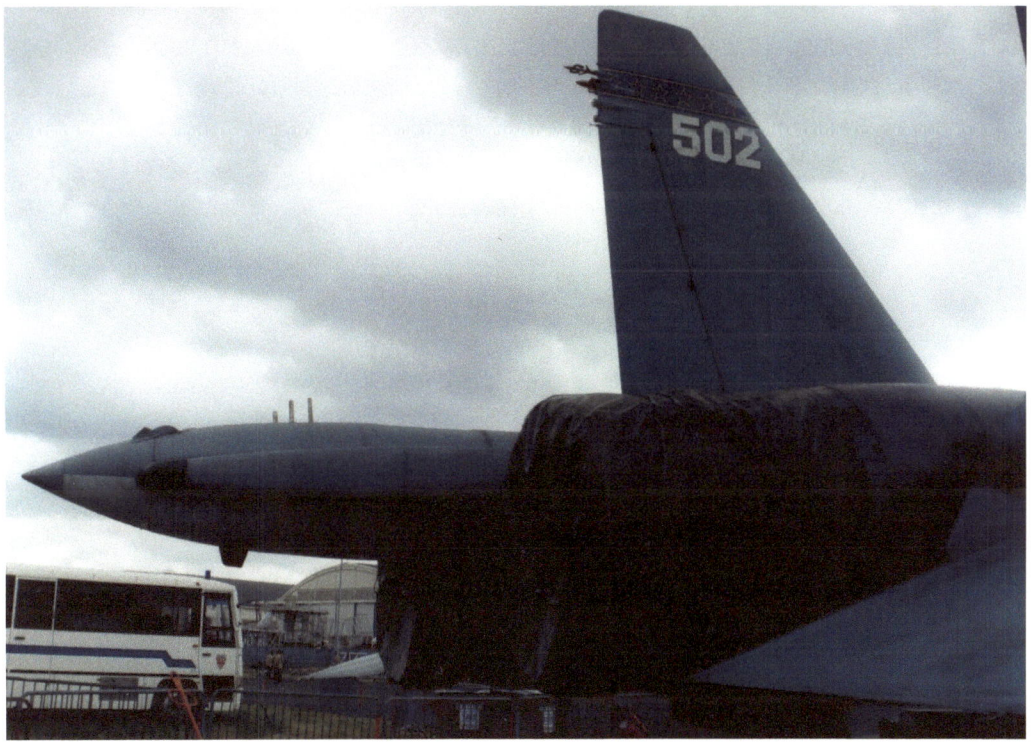

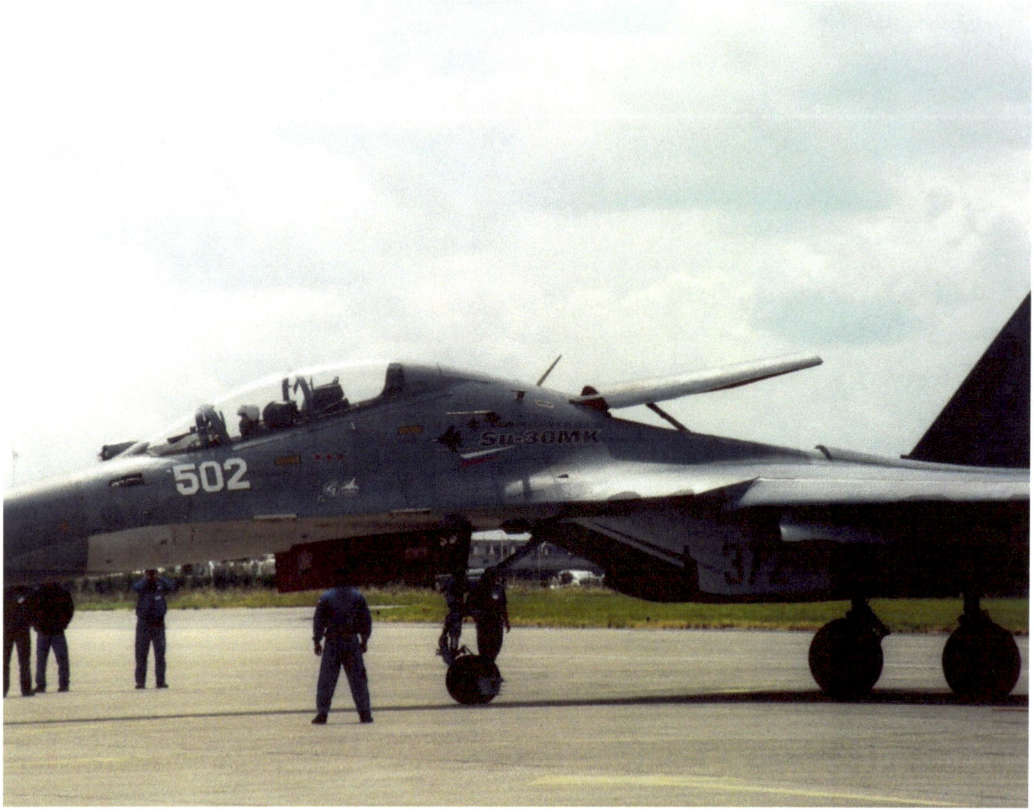
502
Su-30MK

502
Su-30MK
502

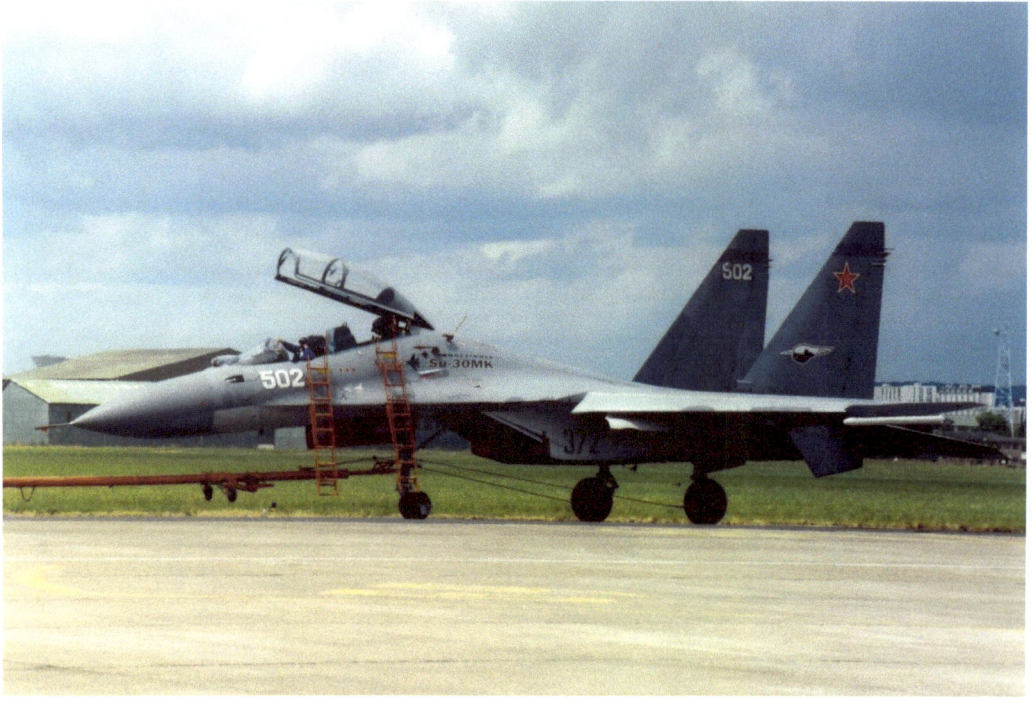

Page 27 top: Su-30MKK starboard wing with leading-edge surfaces in the drooped position. Page 27 bottom: The tail boom appears to be a standard Su-27 family 'sting' incorporating such equipment as the braking parachute and various antennas. Page 28 top: Su-30MKK, White 502, in the process of closing the large slab-type airbrake characteristic of the Su-27/30 family. Bottom: The cockpit is covered by a one piece canopy that hinges at the rear. Page 29-30: Su-30MKK forward fuselage. Author

AL-31F by-pass turbofan engine. NPO and Author **Current performance figures for the AL-31F are: 12500 kgf maximum thrust, 0.695 minimum fuel consumption kg/kgf/per hour, air consumption 112 kg/s. Length is 4.945 m, and dry weight is 1520 kg.** Salut

A defining feature of the Su-30MKK was the fact that it consisted completely of Russian equipment, whereas the Su-30MKI family featured equipment sourced from a number of countries including Russia.

While the Su-27SK/UBK were equipped with the export standard of the N001M radar and OLS-27 (Article 36Sh) Optical Location Station, new variants of these systems were developed for the Su-30MKK Air to Air, Air to Ground Weapon Control System. The SUU-VE WCS (Weapons Control System) consists of the RLPK-27VE (Article N001VE) pulse-Doppler fire control radar, Optical Electronic Sighting System OEPS-27MK (Article 31E-MK), Optical Location System OLS-27MK (Article 52Sh), SURA-K HMTDS (Helmet Mounted Target Designation System), SILS-27M HUD (Heads up Display) and the 6231R- IFF interrogator. Note: some sources state that the HUD is designated SILS-31 and the OEPS-27MK and OLS-27MK are designated OEPS-30 and OLS-30 respectively, however, this appears to be erroneous as Tikhomirov NIIP, which oversees integration of all WCS components, states in its documentation that these systems are SILS-27M, OEPS-27MK and OLS-27MK respectively. To facilitate the delivery of precision guided air to surface weapons an SUV-P sub-system, incorporating 4 x BTsVM-486 computers, is incorporated into the WCS. To the same end, the N001VE is equipped with a Baget reprogrammable digital single processor, an air to surface datalink, an MVK-RL datalink receiver, and a data exchange terminal adapter.

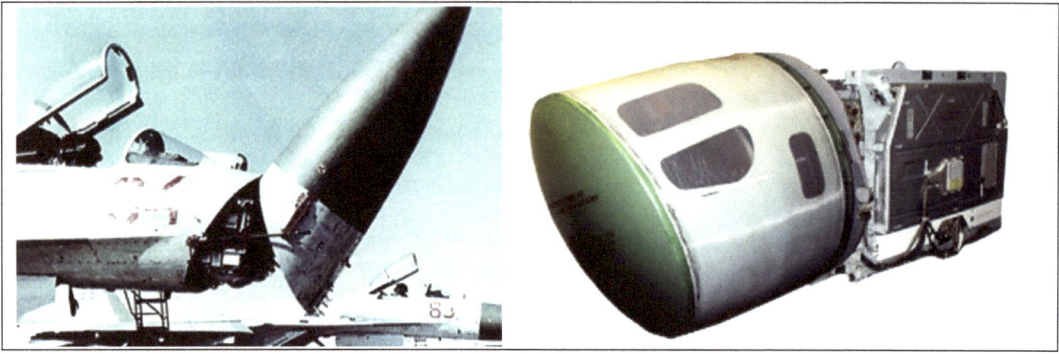

The Su-30MKK, as well as the later Su-30MK2 family, are equipped with enhanced variants of the weapon control system developed for the Su-27. The Su-30MKK is equipped with the N001VE while the Su-30MK2 is equipped with the N001VEP radar. Tikhomirov NIIP

The N001VE incorporates a number of new features including enhanced target designation and "mapping capabilities" and the capability to fire and guide the RVV-AE active radar guided medium range air to air missile. Although the radar standard for the later Su-30MK2 is known in detail, that of the Su-30MKK is more hazy. There are additional modes allowing the radar system to track a reported ten airborne targets. Although it has been widely reported that the version of radar incorporated in the Su-30MKK can be used to engage two targets simultaneously; in either automatic or manual target selection, there is some uncertainty, such a capability being offered as an upgrade for the radar incorporated in the Su-30MK2.

Top: The main elements of the OLS-27MK Optical Location System are located ahead of the windscreen offset to starboard as seen on Su-30MKK White 502 above. The view of the outside world in the frontal aspect is dominated by the SILS-27M Heads Up Display. Author **The SURA-K HMTDS is identical in overall dimensions to the improved SURA-I above.**

An additional mode apparently allows a single target to be engaged by two missiles simultaneously, although it is unclear if this is incorporated on the Su-30MKK N001VE or available only in the N001VEP of the Su-30MK2.

The OLS provides the Su-30MKK with a passive (radar silent – non emitting radar emissions) detection, tracking and engagement capability, reducing the overall vulnerability to enemy direct and indirect detection, tracking and engagement systems and countermeasures. The OLS can also be employed in conjunction with the radar system with which it is completely integrated.

The Su-30MKK/MK2 family is equipped with the OEPS-27MK (Article 31E-MK) incorporating the OLS-27MK (Article 52Sh). Sukhoi

The ECM (Electronic Counter Measures) suite includes an L-150 ELINT system that provides targeting information to the Kh-31P anti-radiation missiles in the defence suppression role. In addition, similar data is supplied to a system apparently known as the Tekon system that is apparently incorporated in the APK-9E pod used to guide the Kh-59ME TV (Television) guided air to surface missile to the target. Respective weapons are launched by data supplied via an armament management system apparently designated SUO-30PK, which was developed by Aviaavmatika.

In the Su-30MKK the two crew are accommodated on K-36D-3.5 zero-zero ejection seats. This type of seat is also installed in the Su-30MK2 variant with the exception of aircraft supplied to Vietnam which are equipped with the K-36D-3.5E model. The manufacturer description reads, "The crewmember protection against the dynamic pressure G-loads at ejection is provided with the protection gear, windblast shield, forced restraint in the seat, seat stabilization as well as the selection of one of three operation modes for the emergency source depending on the suited pilot mass. At the aircraft speed exceeding 850 km/h, the MRM steady-state mode is adjusted by the automatics depending on the acceleration."

"After automatic separation of the pilot from the seat, the recovery parachute canopy is inflated providing the pilot's safe descent. A portable survival kit, which is separated from the seat together with the pilot, supports his/her vital functions after landing or water landing, makes the pilot search easier, and the… -1 life raft supports the pilot floatation on the surface of the water."

"The K-36D-3.5 ejection seat realizes the cremember emergency escape within the range of equivalent airspeed (V_E) from 0 to 1300 km/h, at Mach number up to 2.5 and aircraft flight altitude from 0 to 20000 m, including takeoff, landing run and H-0, V=0 mode. The seat is used with the KKO-15 set of protective gear and oxygen equipment." The weight of the seat and survival kit is around 103 kg.

The pilot is equipped with a SURA-K HMTDS (Helmet Mounted Target Designation System), which, together with the 6231R-9-2 IFF interrogator, completes the WCS. The SURA-K can scan the airspace +/- 60° in azimuth and -20 to +60° in elevation.

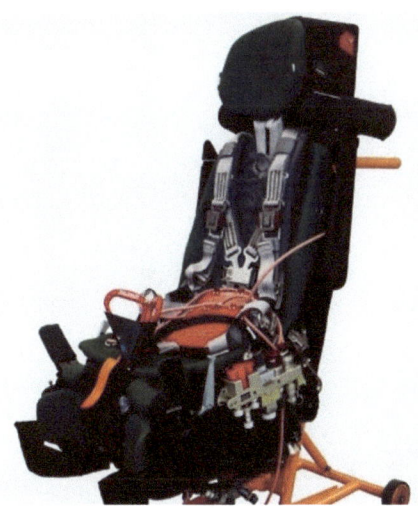

The partial glass cockpit, although appearing outdated by Western standards, is two generations ahead of the archaic cockpits found in previous Chinese fighters of the J-7 and J-8 types and a generation ahead of the cockpit standard of the Su-27SK/UBK. The cockpit, which is dominated by the SILS-27M HUD in front of the windscreen, featured an open architecture avionics suite developed by Ramenskoye RPBK Instrument Manufacturing Design, which included BTsVM-496 computers and two 6 in x 8 in class MFI-10-5 colour MFLCDS (Multi-Function Liquid Crystal Display Screen) in each cockpit. The display screens can replicate all relevant targeting, weapons, fuel, flight and navigational data. Other systems include an A-737 satellite navigation system that is compatible with the GLONASS and NAVSTAR systems and there are improvements to the communications and navigation suites compared to the Su-27SK/UBK.

The integrated self-defence suite includes Sorbtsiya ECM pods, carried on the wingtip stations, threat warning and decoy systems; UV-30MK chaff dispenser, which, as stated by JSC Tactical Missiles Corporation, jam "guidance systems and executive mechanisms with passive countermeasures effective in optical and radar frequency bands".

Top: Zvezda K-36D-3.5 zero-zero ejection seat. Zvezda **Above: Su-30MKK front cockpit (left) and rear cockpit (right).** Sukhoi

| Smart coupling units providing for linking between avionics and weapon | Integrated controller of autonomous operation (IKAR) | Peripheral coupling units providing for linking between avionics and weapon |

Top: Sorbtsiya EW pod on the starboard wingtip station of Su-30MKK White 502. Author **Above: Su-30MKK avionics to weapons interface.** Aviaavmatika

Top: Sorbtsiya EW pod on the starboard wingtip station of Su-30MKK White 502. Author **Above: Su-30MKK avionics to weapons interface.** Aviaavmatika

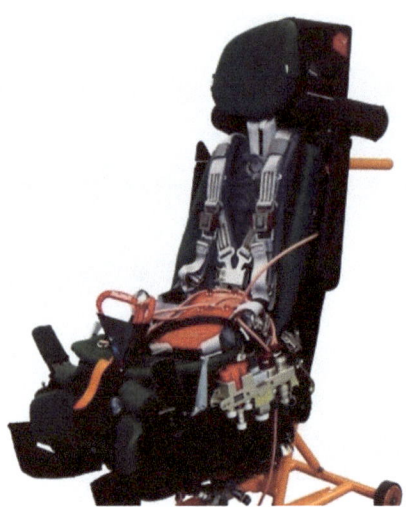

The partial glass cockpit, although appearing outdated by Western standards, is two generations ahead of the archaic cockpits found in previous Chinese fighters of the J-7 and J-8 types and a generation ahead of the cockpit standard of the Su-27SK/UBK. The cockpit, which is dominated by the SILS-27M HUD in front of the windscreen, featured an open architecture avionics suite developed by Ramenskoye RPBK Instrument Manufacturing Design, which included BTsVM-496 computers and two 6 in x 8 in class MFI-10-5 colour MFLCDS (Multi-Function Liquid Crystal Display Screen) in each cockpit. The display screens can replicate all relevant targeting, weapons, fuel, flight and navigational data. Other systems include an A-737 satellite navigation system that is compatible with the GLONASS and NAVSTAR systems and there are improvements to the communications and navigation suites compared to the Su-27SK/UBK.

The integrated self-defence suite includes Sorbtsiya ECM pods, carried on the wingtip stations, threat warning and decoy systems; UV-30MK chaff dispenser, which, as stated by JSC Tactical Missiles Corporation, jam "guidance systems and executive mechanisms with passive countermeasures effective in optical and radar frequency bands".

Top: Zvezda K-36D-3.5 zero-zero ejection seat. Zvezda **Above: Su-30MKK front cockpit (left) and rear cockpit (right).** Sukhoi

The first prototype Su-30MKK, Blue 501 (top), conducted its maiden flight on 20 May 1999, followed that summer by the first flight of Su-30MKK-2, 502, that summer.
Sukhoi

As the first of the four Su-30MKK prototypes commenced manufacture, KnAAPO converted the first Su-30 prototype, T-10PU-5, Blue 05, to test some features of the Su-30MKK, and this aircraft conducted its first flight in support of the MKK program on 9 March 1999, although there are also reports that the first flight in support of the MKK program took place on 5 March that year. This aircraft was little more than a standard Su-30 wearing a new coat of paint along with some minor modifications.

Su-30MKK-1, side code 501 and Su-30MKK-2, side code 502, take-off (top) and fly in formation (above). Sukhoi

The first two Su-30MKK development aircraft were followed by a further two, Su-30MKK-3, side code 503 (top), and Su-30MKK-4, side code 504 (above). Sukhoi

The four Su-30MKK development aircraft were used to verify all aspects of the designs airframe, sensor and avionics performance, as well as weapons integration. Sukhoi

Page 41-45: The second Su-30MKK development aircraft, White 502, at the Paris Air Salon in June 2001. Author

Su-30MKK-2, White 502, taxis at a test airfield during the flight development program. Sukhoi

PLAAF Su-30MKK with various ordnance. KnAAPO

Once the design review was passed construction of the first prototype Su-30MKK commenced at KnAAPO, the aircraft being assembled in spring 1999 before entering a ground test phase prior to its first flight, which occurred on 20 May that year with test pilots I. Ye. Solovyov from Sukhoi Design Bureau and A. V. Pulenko from KnAAPO at the controls. This aircraft, designated Su-30MKK-1 by KnAAPO, received the side code Blue 501. Blue 501 was joined in the test program by another three aircraft, S-30MKK-2, -3 and -4, these aircraft, which received the side codes Blue 502, 503 and 504 respectively, being referred to by KnAAPO as preproduction aircraft.

When the PLAAF started taking delivery of its first batch of Su-30MKK's these aircraft were the most advanced combat aircraft in that nation's inventory. The PLAAF eventually received seventy six Su-30MKK's, which were ordered in two batches, each of thirty eight aircraft. PLAAF

The second Su-30MKK, Blue 502, was completed in summer 1999, flying a short time later. Unlike the previous examples the next two Su-30MKK's, Blue 503 and Blue 504 took to the air in yellow primer with which they continued to be adorned for some time during flight testing. Once completed the four development (pre-production) Su-30MKK's were delivered to Sukhoi Design Bureau for testing between 1999 and 2001 in cooperation with SPFC of the Russian Federation Air Force, testing at the GLIT Chkalov, Russian Federation Air Force State Flight Test Centre, apparently being completed towards the end of 2000.

The most visible of the Su-30MKK development aircraft was the second pre-production aircraft, code 502, which was seen at Air show China 2000 at Zhuhai-Sanzao airport, that time as White 502, which it still retained at the Paris Le Bourget show in June 2001.

A Su-30MKK for the PLAAF. KnAAPO

The anticipated order for the Su-30MKK from China became a reality when the contract for same was signed on 27 August 1999; thirty eight aircraft being ordered, just as the Su-30MKK flight test program was ramping up in Russia.

The first batch of ten production standard Su-30MKK's were delivered to China on 20 December 2000 (KnAAPO documentation indicates that ten aircraft were delivered to the customer in December 2000, however, other sources, unconfirmed, suggest that only six Su-30MKK's had been received by China by the end of December 2000, the balance of four being received in January 2001). The aircraft delivered on 20 December conducted a ferry flight from KnAAPO to Wuhu, China, where they were incorporated into the PLAAF 3rd Air Division. Around the same time China also received a batch of four Su-27UBK two-seat operational trainers. The balance of the order for thirty eight Su-30MKK's was delivered during the course of 2001, this including a batch of eight aircraft apparently delivered in April that year.

The reported inauguration of the Su-30MKK into PLAAF service is claimed to have taken place on 20 January 2001, and the aircraft were establishing themselves in service, when in December 2001, China ordered a second batch of thirty eight Su-30MKK's for the PLAAF. A sub-batch of this order, apparently numbering ten aircraft, was delivered in August 2002, apparently the 19th, a further batch of nine aircraft being delivered in December 2002, apparently the 16th. These aircraft were apparently flown to a PLAAF air base in Tsuizhou in company with a pair of Ilyushin Il-76 transport aircraft which were carrying support equipment and spare parts. The balance of nineteen aircraft from the 2001 order was delivered during the course of 2003, bringing to seventy six the number of Su-30MKK's delivered to China.

While confirmation of operational units within the PLAAF remains difficult, as far as can be ascertained Su-30MKK's were, by 2006-2007, serving with the 3rd Air Division (apparently the 9th Regiment), Wuhu, Nanjing Military Region, the 18th Air Division at Datuopu located in Changasha, Guangzhou Military Region and the 29th Air Division (apparently the U/I Regiment) at Quzhou, Nanjing Military Region. In addition, a number were operated by the PLAAF Test and Training Centre located at Cangzhou.

The standard Su-30MKK weapons capability consisted of the internal cannon, RVV-AE medium range active radar guided air to air missiles, R-27R1 medium range semi-active radar guided air to air missiles, R-73E short range infrared guided air to air missiles, Kh-29L/T/TE, short range air to surface missiles, Kh-31P anti-radiation missiles, Kh-59ME (with associated guidance pod) long-range TV guided air to surface missiles, KAB-500 and KAB-1500 guided bomb units as well as a diversity of unguided bombs and rockets.

China apparently purchased 100 RVV-AE missiles on 10 May 2000 for service with the designation of R-129. In summer 2002, Su-30MKK's conducted a number of training flights during which it is claimed a number of such missiles were launched.

4

IMPROVING THE BREED – Su-30MK2/M2

An enhanced evolution of the Su-30MKK was developed for China's PLNAF (Peoples Liberation Naval Air Force). This new variant, which differs from the Su-30MKK only in the equipment fit and weapons suite, went on to form the basis of a standard export model under the designation Su-30MK2, being purchased not only by China, but also Indonesia, Vietnam and Venezuela. More recently the Su-30M2 domesticated variant of the Su-30MK2 has been purchased in modest numbers by the Russian Federation Air Force along with larger numbers of the more capable Su-30SM super-manoeuvrable multi-role fighter which is built at Irkut.

The main difference between the Su-30MKK and Su-30MK2 is in the standard of the Weapons Control System, which, in the Su-30MK2, is designated SUU-VEP, This consists of the RLPK-27VEP (Article N001VEP) pulse-Doppler fire control radar, OEPS-27MK (Article 31E-MK) Optical Electronic Sighting System, OLS-27MK (Article 52Sh) Optical Location System, SURA-K HMTDS (Helmet Mounted Target Designation System), SILS-27M HUD (Heads up Display) and the 6231R-IFF interrogator.

The manufacturer, V. Tikhomirov NIIP, description of the system reads, "…weapon control system WCS-VEP (SUU-VEP) is intended for air target search, identification and aiming at collision courses and in Tail hemisphere in Look-up and Look-down modes in overwater and overland environment. The given WCS is mounted on the aircraft of Su-30MK2, Su-27SM types intended to achieve the air superiority, to hit ground and surface targets by means of guided and unguided weapons while group or single actions day-and-night under good or bad weather as well as to fulfill long-range patrolling and tracking." In regards to application, the manufacturer continues, "During combat employment of Su-30MK2 aircraft WCS provides for solution of the following tasks: - turning the aircraft to the area of the assigned target; search, detection, identification, lock-on, auto-tracking, coordinate and parameter measurement of air targets; aiming, generation of target designation &

weapon and aircraft control commands and signals; air target hitting by means of guided missiles with RHH and IRHH and built-in guns; en-route flight and turning to programmed ground target in the given direction."

The pinnacle of the Su-30MKK design, the Su-30M2 has been built for the Russian Federation Air Force in modest numbers with deliveries commencing in 2011. Sukhoi

The following data, provided by Tikhomirov NIIP, pertains to the standard of WCS fitted in the Su-30MK2:

"'Air-to-Air', 'Air-to-Ground' weapon control system WCS-VEP (Sh101VEP) provides for application of the following weapons: RVV-AE, R-27ER1, R-27R1, R-27ET1, R-27T1, R-27EP1, R-27P1, R-73E, X-31A, X-59MK, X-35E, air bombs, unguided missiles, GSh-301."

"Radar aiming complex includes: RLPK-27VEP* * = upgraded ((N001VEP) for the following weapon application: R-27P1, R-27EP1, X-59MK, X-35E."

"On-board digital computer (BTsVM-900), duel channel digital receiver – N001-03VP2 (under upgrade)"

"Digital processor – Baget 55-04.02."

"Intermediate frequency signal switchboard – N001-39 (under upgrade)."

"master oscillator N001-22P."

"bus adapter switch N001-04M."

"Radar type Pulse Doppler."

"Pulse repetition frequency, high, medium, low."

"In 'Air to Air' mode the radar provides as follows: velocity search; search with ranging; air target illumination and transmission of radiocorrection commands to control missiles with radar homing heads; to control missiles with infrared homing heads; search, lock-on and tracking of visually visible target in close combat; target IFF; operation in adversary EW environment; jammer coordinate measurement; interaction with ECM equipment."

"Number of targets with their coordinate measurement in TWS mode, pcs, 10."

"Number of simultaneously attacked targets, 1* (*can be increased to 2)"

"Detection and tracking zone": +/- 60° in azimuth and -55° - +60° in elevation."

"Search and lock-on zone in close combat": +/- 2° in azimuth and -10° - +50° in elevation.

"Detection range for an air target of a fighter type (RCS=3m^2, with 0.5 probability)", look-up in forward hemisphere – no less than 100 km (in long range detection mode the range can be increased to 150 km); look-up tail hemisphere – no less than 40 km; look-down in forward hemisphere – no less than 80 km; look-down in tail hemisphere – no less than 35 km.

"Operation range of RVV-AE radio correction channel: up to 40 km".

"In 'Air to Ground' mode the radar provides as follows: detection of ground and surface targets in real beam mapping mode, while scanning in low resolution mode (LRM), detection of ground and surface targets in SAR mapping mode in medium and high resolution modes (MRH, HRH), detection of ground and surface moving targets in ground moving target selection mode (GMTS), tracking and coordinate measurement for a ground target, output of target designations to X-31A, X-59MK, X-35E missile RHH."

"In GMTS mode the radar provides for the detection of moving targets with RCS of about 10m^2 (a tank) and more, and radial velocity" at ranges of 15-90 km.

"Characteristics in 'Air to Ground' mode, limits of search zone size: mapping in Real Beam mode (RB) +/- 45° (within +/- 60° angles); in sea search mode (SS) +/- 45° (within +/- 60° angles); in Doppler beam sharpening (DBS) 30° within +/- (10° ...60°) angles; in HRH mode 5° within +/- 30° ...60°) angles."

"Detection range" is stated as "'350 km' for an aircraft carrier with an RCS of 50000 m^2, '250 km' for a Destroyer with a RCS of 10000 m^2, 'no less than 100 km' for a railway bridge with RCS of 2000 m^2, '50-70 km' for a missile boat with a RCS of 500 m^2 and '30 km' for a boat with a RCS of 50 m^2."

The Su-30MK2 is equipped with an OEPS-27MK (Item 31E-MK) Optical Electronic Sight System which also "(provides for a ground target illumination)".

The Optical Location System element of the OEPS system consists of the "OLS-27MK (Article 52Sh)" with an in azimuth tracking zone of "+/- 60°" and "-15 ...+60°" in elevation.

The system features a "60°" in azimuth and "10°" in elevation field of view and search; a "20°" in azimuth and "5°" in elevation "small field of view and search"; "Close combat area ('Vertical' mode) 3 x (-15° ...+ 60°; Lock-on area 3° x 3°".

The system has a detection range against an air target in infrared contrast in the tail on aspect for a target in the class of a Sukhoi Su-15 interceptor operating without afterburner "(PMFU)" of "no less than 30 km". A target in the class of the RAC MiG-25 at high altitude in the forward aspect in afterburner at speeds of at least Mach 2.0 can be detected at "no less than 90 km".

Against an airborne target in the RAC MiG-21 class the laser rangefinder can be operated at ranges up to "8 km" and "0.3 …10 km" against ground targets.

The other elements of the WCS, as noted above, consists of the "SILS-27M" HUD (Heads up Display), "6231R-9-2" IFF interrogator and SURA-K Helmet Mounted Target Designation System which can scan the airspace "+/- 60°" in azimuth and "-20 to +60°" in elevation.

Often erroneously stated as being a standard feature of the Su-30MK2 WCS, the radar is capable of being enhanced to allow engagement of two targets simultaneously. The WCS was designed to take various upgrades including integration with new advanced air to air and air to surface weapons – RVV-MD close-range infrared guided, R-27P(EP1) passive radiation homing medium-range and RVV-SD active-radar homing medium range air to air missiles; precision guided bombs with satellite guidance, KAB-500C-E (KAB-500S-E). "Provision for new weapons application (470OUT-RT, X-31P-UL, X-31A-UL) to improve the quality of aviation specialist training."

Su-30MK2 Serial production of the Su-30MK2 multipurpose fighter for export deliveries

The Su-30MK2 was exported to China, Indonesia, Vietnam and Venezuela and Russia purchased a small number of a domesticated variant designated Su-30M2. KNAAZ

As with the Su-30MKK, the cockpit of the Su-30MK2 is dominated by the SILS-27M HUD (forward cockpit only) and 4 MFDS, two in each cockpit station, these forming the major elements of the cockpit management system. As with the Su-30MKK, the crew of two are accommodated on standard K-36D-3.5 zero-zero ejection seats with the exception of aircraft supplied to Vietnam which are equipped with the K-36D-3.5E seat.

The significant addition to the weapons suite was the Kh-31A anti-ship missile, allowing the aircraft to take on an enhanced maritime strike role. The Su-30M2 variant in service with the Russian Federation Air Force has a broader weapons capability than the proceeding export variants.

Any additional flight development work for the MK2 program was conducted by the MKK test fleet. Author

There are conflicting dates for the finalisation of the contract signing for the purchase of twenty four Su-30MK2's for the PLNAF; these dates are 27 August 2002 and January 2003 respectively. Likewise, there are conflicting dates for the first deliveries to China of a less than precise 2003 (presumably late in the year) and the more precise February 2004. What is clear is that all twenty four aircraft had been delivered before the end of 2004. As with the PLAAF, operators are hard to ascertain for certain, but appear to be the East Sea Fleet 4th Air Division located at Ningbo.

By the end of 2004, China had acquired 76 Su-30MKK's and 24 Su-30MK2's. By then China had also delivered or received assembly kits for 105 licence produced Su-27SK's on top of the 76 Su-27SK/UBK's delivered from Russia, elevating China to the position of second largest operator of Su-27 and derivatives.

In the first decade of the 21st century the Su-30MK2 formed the basis of KnAAPO's export drive, achieving sales in three separate continents in as many years. Indonesia required a new advanced multi-role fighter aircraft to supplement and eventually supplant the TNI (Indonesian Military) small fleet of Lockheed Martin (formerly General Dynamics) F-16A/B Fighting Falcons, which were suffering from obsolescence (with regards to avionics and weapons) and operational capability due to the difficulty in obtaining spares in the face of a US arms embargo, prior to which the US supplied around 70% of Indonesia's military hardware.

The Su-30MK2 was initially developed to meet the PLNAF requirement for an enhanced maritime air to surface strike capability.

Although the planned purchase of the single-seat Su-30KI did not materialise due to financial constraints, Indonesia retained its requirement for modern fighter aircraft leading to small scale purchases of Su-27 derivatives from 2002. The numbers of aircraft purchased constituted only a token force of two Su-27SK and two Su-30MK2's. The Su-27SK's were delivered as air freight on 27 August 2002, followed by delivery of the two Su-30MK2's on 1 September 2003; the first of the two Su-30MK2's received the serial TS001 and fuselage side code Red 01.

Following delivery the Su-30's were assembled by Sukhoi engineers with flight testing commencing on 11 September 2003. As deliveries took place a detachment of six air crew and eighteen ground-crew was training in Russia in preparation for Indonesian air force operations. The Su-30's were delivered to Iswahyudi air base in Madiun district for service with 11 Skvadron (Squadron), and the aircraft were displayed at the TNI anniversary event on 5 October 2003.

The Indonesian aircraft were delivered without some systems, which, combined with the token number purchased, reduced their operational potential. In the longer term the Indonesian military hoped to increase the number of Su-27/30's to between 12 and 16 fighters, which would allow equipment of a full operational squadron mix of Su-27SK/SKM and Su-30MK2's.

The introduction of the Su-30MK did give the Indonesian Air Force a token ability to fly to targets as far away as Malaysia's Sarawak state as well as allowing it to cover the Indonesian provinces of Central Sulawesi, East Nusa Tenggara and Jambi as well as covering much larger areas of Australia than was the case with the smaller F-16.

In late 2003 early 2004, it was announced that Indonesia hoped to purchase a further six Su-27SK and a pair of Su-30MK2's. As part of the drive for further Su-30 sales to Indonesia, Sukhoi took a Su-30MK2 to INDO-Defence 2006 held in Indonesia in November 2006. At the show a further Su-30MK2 sale was officially offered to Indonesia and Sukhoi revealed that talks on an Indonesian purchase of a batch of Su-30MK's were then underway leading to a contract being signed for the purchase of three Su-30MK2's and three Su-27SKM single-seat fighters on 23 July 2008, facilitated by a Russian Federation loan to Indonesia. Two of the Su-30MK2's arrived in Indonesia from KnAAPO on 29 December 2008, and these aircraft were flight tested on 6 January 2009. On 17 January 2009, the third Su-30MK2 was delivered to Indonesia and was assembled and flight tested at Sultan Hasanuddin air base (airport), Sulawesi, over the next few days; the flight lasted one hour and was conducted by a Russian crew; all of the aircraft on-board systems being tested. The three Su-30MK2's were officially handed over to the Indonesian Government on 2 February 2009.

Indonesian Su-30MK2 TS001, Red 01. KnAAPO

The three Su-27SKM fighters were delivered in September 2010; two being flown from KnAAPO to Sultan Hasanuddin air base, where they arrived on the 10th of the month, followed by the third which was delivered by transport aircraft on the 16th of the month. These six new fighter aircraft, together with the two older Su-27SK and two Su-30MK2's purchased in 2003, would form an operational Squadron.

Supplementing its meagre fleet of twelve Su-27 fighters purchased in the 1990's, Vietnam ordered four Su-30MK2's in 2003, these being delivered in November 2004. These aircraft were more or less to the same standard as those aircraft delivered to the PLNAF, the only differences apparently being slight modifications to the ejection-seats and communications equipment. Although often referred to as Su-30MKV, the aircraft are designated Su-30MK2 by KnAAPO and simply as Su-

30MK by Sukhoi. As with the Su-27's, the Su-30MK2's were based at Phan Rang air base and operated by the 370th Air Division.

Vietnam went on to order an additional batch of eight Su-30MK2's, the last of which were delivered in 2010. A further batch of twelve Su-30MK2's was ordered, apparently in January 2010, with deliveries from late that year and into 2011, this bringing to twenty four the number of Su-30MK2's operated by Vietnam.

In the early 2000's, the Venezuelan Air Force was looking for a replacement for the services fleet of obsolescent F-16A/B Fighting Falcon fighters. The need for an advanced fighter to replace the F-16 stemmed from what the Venezuelan President perceived as an increasingly belligerent approach to that nation by the United States; to the point where Venezuela feared direct offensive or indirect action by the fermentation of internal unrest to destabilise the oil rich nation with the aim to overthrowing the Venezuelan government.

Having tried in vain to obtain spares for its F-16 fleet from the United States, Venezuela decided to replace the US aircraft and looked at a number of options from Russia, including the MiG-29, Su-27 and Su-30. In July 2006, a pair of KnAAPO Su-30MKK development aircraft, 501 and 502, arrived in Venezuela and participated in the celebrations of the country's independence from Spain. The Su-30MKK's, by then referred to as MK2 development aircraft, flew over 14000 km to Venezuela from Vnukovo airport, Moscow to El Libertador air base, Central Aragua, Venezuela, with stopovers at Varna (Bulgaria), Malta, Casablanca (Morocco), Sal Island (Cape, Verde), Recife (Brazil) and Paramaribo (Surinam). The journey taking place between 29 June and 2 July; the aircraft being flown by Sukhoi test pilots Sergey Bogdan and Vyacheslac Aeryanov.

The high profile display of the Su-30's came at the same time as the Venezuelan president announced that nations intention to order twenty four Su-30's, the contract for the purchase being signed during the Venezuelan President's visit to Moscow on 25 July 2006. The $1 billion contract covered the purchase of twenty four Su-30MK2 strike fighters, adding to an earlier contract for the purchase of 33 helicopters; 10 Mi-35M, 20 Mi-17V-5 and 3 Mi-26T, to equip a rapid reaction battalion.

Venezuela's requirement to have the first of the Su-30MK2's delivered in time for the Venezuelan Air Force Day on 10 December 2006 was a tight schedule. Following assembly the first of the aircraft entered flight testing in October 2006, and by the end of that month they had been painted; one aircraft being adorned in a special colour scheme that included the Venezuelan Flag on the tail fins. A delegation from Venezuela officially accepted the first two aircraft at KnAAPO in early November. The two Su-30MK2's were then prepared for delivery to Venezuela, being transported on-board an An-124 Ruslan transport aircraft to Luis del Valle Garcia air base, Barcelona, some 230 or so km to the east of the Capital Caracas, where they arrived on 1 December. After delivery the aircraft were assembled and test flown by Russian test pilots Alexander Pulenko and Pavel Tarakanov before being handed over for acceptance by the Venezuelan Air Force with aircraft side codes 1259 and 0460. On 6 December 2006, the two aircraft were flown by Russian pilots as an escort for the Venezuelan Presidential aircraft, Airbus A319CJ, 0001.

Page 59-60: The first two Su-30MK2's for the Venezuelan air force during tests at **KnAAPO.** Sukhoi

The schedule having been met, the Venezuelan Air Force unveiled the first two Su-30MK2's's at the display held at the Venezuelan Air Force Day (86th anniversary of Venezuelan aviation) at El Libertador air base on 10 December 2006, with the Venezuelan President in attendance; the aircraft being flown by Sukhoi test pilots Sergey Bogdan and Vyacheslav Averyanov with Venezuelan Air Force pilots in the rear seats.

Two more Su-30MK2's were delivered to Venezuela around the middle of December 2006. Another eight Su-30MK2's had been delivered by the end of September 2007, with several others delivered by the end of that year and the remaining aircraft were delivered in 2008. The Su-30MK2's were allocated to the reactivated 13th Air Group based at Barcelona, Eastern Anzoategui State.

Venezuela was widely expected to order additional Sukhoi multi-role fighters with attention switching to the more advanced Su-35S 4th++ Generation thrust-vector control 'super-manoeuvrable' multi-role fighter, which entered service with the Russian Federation Air Force in 2014. However, as of 2015 no order has been announced.

When the Su-30MK2 program emerged in the early 2000's, Russia's economic situation ruled out any domestic purchase of the aircraft. However, towards the end of the first decade of the 21st century the improving economic conditions in Russia led to the first of two orders for a variant designated Su-30M2 for service with the Russian Federation Air Force with a primary air to surface role. KnAAPO

Page 62-64: The Su-30M2 first flight took place at Komsomolsk-on-Amur on 17 September 2010. Sukhoi

The Su-30M2 was developed to meet the Russian Federation Air Force requirement for a modern multi-role strike fighter optimised for the air to surface role, but retaining full air to air capability. For its primary strike role the Su-30M2 can employ a vide diversity of precision guided air to surface munitions as well as unguided bombs and rockets.

The Russian Defence Ministry ordered four Su-30M2 aircraft at MAKS 2009 in August 2009, along with twelve new build Su-27SM3 and 48 Su-35S 'super-manoeuvrable' multi-role fighter aircraft. The first Su-30M2 conducted its maiden flight from KnAAPO on 17 September 2010. KnAAPO conducted a series of preliminary flight tests during September, following which the aircraft was handed over for the certification test phase.

The token Su-30M2 force was augmented by a further sixteen aircraft ordered in December 2012; the four aircraft ordered in August 2009 had been delivered to Russian Federation Air Force units in the Southern and Eastern Military Districts the previous year. Four of the sixteen aircraft ordered in 2012 were delivered in December 2013; these apparently being allocated to three separate air force units in the Southern and Eastern Military Districts. Another four Su-30M2's were handed over on 5 August 2014; these aircraft being allocated to air force units in the Southern Military District. Two more Su-30M2's were delivered on 10 October 2014 and by November that year sixteen of the 20 Su-30MK2's on order had been delivered, the remaining four following over the ensuing months.

There have been no further Su-30M2 deliveries, the Russian Federation Air Force now receiving more advanced, Irkut built, Su-30SM 'super-manoeuvrable' multi-role fighters derived from the Indian Air Force Su-30MKI, but with Russian equipment replacing foreign systems.

Page 65-70: Russian Federation Air Force Su-30M2 painting and roll-out. Sukhoi

Page 71 top: Russian Federation Air Force Su-30M2 with airbrake deployed during landing. Page 71 bottom - page 74: Su-30M2 Red 504. Knaaz

5

OFFENSIVE AND DEFENSIVE WEAPON SYSTEMS

The Su-30MKK/MK2/M2 family can be armed with a wide diversity of air to air and air to surface weapons. There are slight differences in the armament suite of the respective operators of the various variants. For example, the Su-30MKK operated by the PLAAF can be armed with RVV-AE, R-27R1 and R-73E air to air missiles, Kh-29L/T/TE, Kh-31P and Kh-59ME air to surface missiles and KAB-500 and KAB-1500 guided bomb units. With the Su-30MK2 operated by the PLNAF the Kh-31A anti-ship missile is added to the armoury. With both variants other ordnance can be integrated. The Su-30M2 in service with the Russian Federation Air Force is assumed to be capable of operating with the full spectrum of air to air and air to surface weapons operated by other multi-role Su-27/30 variants in service with the Russian Federation Air Force. In the air to air role this would add a number of R-27 variants whilst for the surface strike mission the aircraft can add a number of advanced air to surface missiles such as the Kh-35E/UE and Kh-59MK. All operators can also employ a diversity of unguided air to surface weapons.

In addition to externally carried ordnance all variants are armed with an internal GSh-301 30-mm cannon found in other members of the Su-27 family. This powerful weapon, housed in the starboard wing-root with 150 round of ammunition, can fire at a rate of between 1,500 and 1,800 rounds per minute, with a muzzle velocity of 870 meters per second. The cannon has a range out to around 1800 meters in the air to air role or up to 800 meters against surface targets. For air to air and air to ground missions the cannon is primarily a secondary weapon.

The Su-30MK2/M2 variants can be armed with the Kh-31A anti-ship missile and the Kh-31P anti-radiation missile while the Su-30MKK can be armed with the latter missile. A total of six Kh-31's can be carried, either all of the same variant or a mix of both variants. In the defence suppression role the six Kh-31P anti-radiation missile are carried on the following stations; one on each of the intake stations, one on each of the inner wing stations and one on each of the intermediate wing stations.

Su-30MKK with a diversity of ordnance including Kh-31 and Kh-59ME guided missiles and KAB-500/1500 guided bomb units. Author

Documentation provided by JSC Tactical Missiles Corporation shows that the Kh-31P features "changeable passive radar homing heads… operating in corresponding frequency bands" allowing it to engage "modern continuous-wave and pulsed radar" systems employed by medium and long range surface to air missile systems. The missile can also engage other emitting radar systems not necessarily part of the air defence system. The homing head autonomously searches for and locks-on to a target, or, alternatively the launch aircraft sensors can hand down targeting information to the missile before it is launched from the AKU-58 airborne ejection unit.

A modified variant of the missile, designated Kh-31PK, employs a larger warhead that is detonated by a proximity fuse. This variant retains the same operating parameters to those of the Kh-31P. The Kh-31PD is an evolution of the Kh-31P, range being increased from a maximum of 110 km to 250 km whilst carrying a more powerful warhead.

The Kh-31P, which has a launch weight of around 600 kg, is 4.7 m in length, 0.36 m in diameter and has a wing span of 0.914 m. The missile can be launched from altitudes of 100-15000 m at a carrier speed of Mach 0.65-Mach 1.25, after which it flies to targets between 15-110 km away (depending upon launch altitude) at speeds of 1000 m/s. The target is destroyed by an 87 kg high explosive fragmentation warhead.

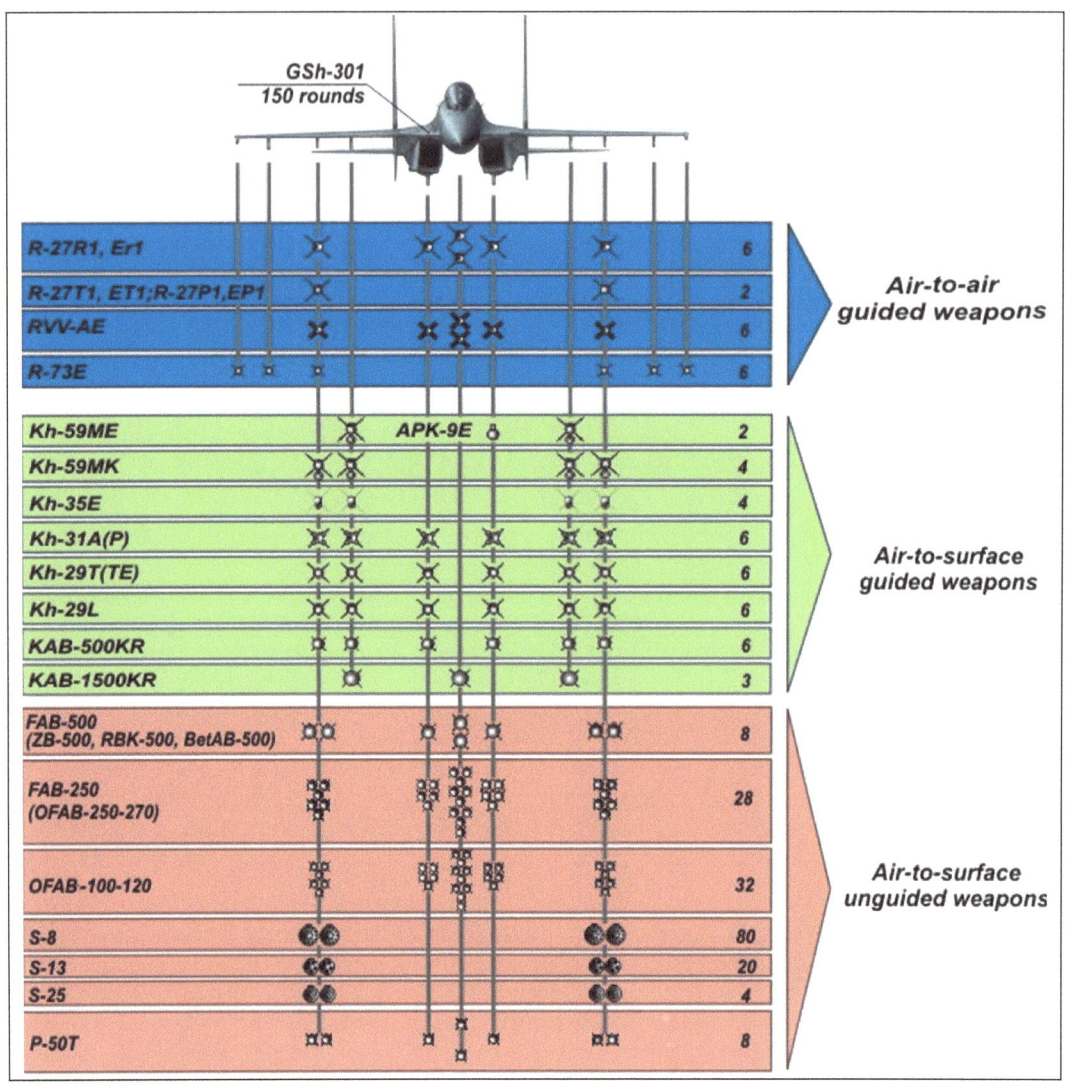

Page 77: Grainy stills from GSh-301 30-mm cannon trials on the Su-30MKK prototype, 501. Sukhoi

Above: Chart depicting the standard air to air and air to surface weapons load out for the Su-30MK2. In the air to surface mission the aircraft can operate with a diverse mix of guided and unguided air to surface stores. In the air to air role the Su-30MK2/M2 variants can engage targets with active radar guided RVV-AE, semi-active-radar guided R-27ER1/R1, radiation homing R-27EP1/P1 and infrared guided R-27ET1/T1 medium range missiles, as well as the highly agile R-73E infrared guided short-range air to air missile. Sukhoi

While only the Kh-31P anti-radiation variant is integrated with the Su-30MKK, the Su-30MK2 adds the Kh-31A anti-ship missile to the armoury. This Kh-31 mock-up (left) is shown with an R-27 semi-active radar guided air to air missile. Author

Kh-31A

As with the Kh-31P, the Su-30MK2/M2 can carry up to six Kh-31A anti-ship missiles. Developed as a high-speed air launched anti-ship missile, the Kh-31A is designed to engage warships operating independently or as part of a larger integrated naval group. The missile, which has the same overall dimensions, similar launch weight, and identical launch parameters as the Kh-31P, can be launched from the carrier aircraft singly or "in salvo" in clear and adverse weather conditions, against background clutter in an active jamming environment. The missiles on-board active-radar homing head can designate targets in both pre-and-post launch modes and conduct target acquisition and selection, and, according to manufacturer documentation, determines "target coordinates (range, azimuth, elevation), generation of command signals", which are fed directly to the guidance system. The missile is carried on and launched from the AKU-58A ejection unit, cruising at a speed of 1000 m/s to targets 5-70 km distant (against a Destroyer size target) depending on launch altitude. The target is then destroyed or disabled by the 95 kg warhead.

The Kh-31AD is an evolution of the Kh-31A with many improvements including a 15% more powerful warhead and longer range; the latter being more than twice that of the Kh-31A.

A Kh-31P anti-radiation missile is launched from the fourth Su-30MKK development aircraft, 504, during integration testing of the weapon for the PLAAF Su-30MKK.
Sukhoi

Another anti-ship missile integrated with the Su-30MK2 is the Kh-35E, four of which can be carried; one on each of the inner and intermediate wing stations. This weapon, which is designed to destroy surface vessels, including warships displacing up to 5,000 tonnes, can be launched from warships (Uran-E ship-borne missile system), coastal missile batteries (Bal-E mobile coastal launch system) and aircraft launched. The aircraft launched missile has a length of 3.85 m, diameter 0.42 m, wing span 1.33 m and a launch weight of 520 kg.

Once launched from the mother aircraft, with maximum turn angle in horizontal plane after launch of +/- 90°, the missile, which cruises at Mach 0.8, descends to an altitude of some 10-15 m above the sea surface, dropping to 4 m for the terminal phase of the flight, to strike targets up to 130 km distant in sea states up to 6 in an active electronic countermeasures environment; the ARGS-35E active radar seeker having an acquisition range of around 20 km, thereafter the target is locked-on and destroyed or disabled by the 145 kg high explosive penetrator warhead.

The Kh-35UE improves on the Kh-35E in a number of areas, including range, which is doubled from 130 km to 260 km, and features an improved post-launch horizontal turn capability.

Kh-35E (X-35E) anti-ship missile. Author

Kh-59ME

The Kh-59ME is an evolution of the Kh-59M (AS-13 'Kingbolt') introduced in the early 1990's. The missile (technical information pertains to the Kh-59M2E missile), which has a casing length of 5.7 m, casing diameter 0.38 m, wing span 1.3 m and a launch weight of "up to 960 kg", can be launched from carrier aircraft flying at speeds of 600-1000 km/h and from altitudes of 0.2-5 km, the missile flying to targets out to 115-140 km (depending on launch parameters), cruising at Mach 0.72-Mach 0.88 at cruise altitudes of 0.007 (over sea) or 0.05-1 km over land.

Model of a Su-30MKK wearing the 502 code of the second development aircraft. The stores configuration consists of four R-73E air to air missiles on the wingtip and outer wing stations, two RVV-AE air to air missiles on the intermediate wing stations, a KAB-1500 guided bomb on the centre station, an Kh-31P on the starboard intake station, a pair of Kh-59ME air to surface missiles on the inner wing stations and the APK-9 Ovod guidance system on the port intake station.

The Tactical Missiles Corporation description of the system reads "aircraft guided missile X-59M2E with translational command aiming system of the 'Ovod-ME' missile system provides hitting wide range of the ground and surface fixed targets with well known coordinates under conditions of limited visibility including night time."

The control system, as stated in manufacturer documentation, consists of an "aiming and automatic control system on the basis of inertial system unit + uncontrolled emergency jettison + low level television (imaging infrared)". The missile has a stated accuracy of 2 to 3 m in manual mode and 5 to 7 m in automatic mode, the target being destroyed by the penetrating warhead with weights of "320" or "283" kg.

This weapon is also employed by the Sukhoi Su-24M2 strike aircraft, guided by an APK-9 Ovod targeting pod carried by the launch aircraft, which has a datalink to the missile. Prior to missile launch, target coordinates are downloaded for the inertial guidance phase via a datalink. The Kh-59ME is also apparently integrated on the Su-34 'Fullback' intermediate range strike aircraft developed as a replacement for the Su-24 in Russian Federation Air Force service.

A Kh-59 air to surface missile is dropped from the third Su-30MKK development aircraft, code 503, during integration testing of the weapon. The Kh-59ME bestows upon the launch aircraft a long-range precision guided capability against fixed location targets such as buildings and bridges. The missile is guided to the target by the APK-9 Ovod pod mounted targeting system which sends commands to the missile via a datalink. Sukhoi

Kh-59MK

Released Sukhoi documentation shows that the Kh-59MK (AS-18 'Kazoo') is the standard long-range anti-ship strike weapon specified for the Su-30MK2, while the Kh-59MK2 may be integrated for use against land targets. Four of these missiles can be carried; one on each of the inner wing stations and one on each of the intermediate wing stations, whereas only two of the Kh-59ME TV guided air to surface missile can be carried; one on each of the inner wing stations with an APK-9E guidance pod carried on the port engine intake station; the overall system known as the Ovod-ME as noted above.

Raduga unveiled the Kh-59MK air to surface missile, a development of the Kh-59M air to surface missile, at the MAKS 2001 show in August 2001. The Kh-59MK, which has a length of 5.7 m, main body diameter (minus engine) of 0.38 m increasing to 0.42 m at the nose, wing span of 1.3 m and a launch weight of "not more than 930 kg", features an ARGS-58E active radar homing head claimed to be capable of detecting a destroyer size maritime target at a range of 25 km. The missile is powered by a low specific fuel consumption NPO Saturn 36MT turbofan engine, which extends the missiles range to 285 km when employed against a Destroyer size target, reducing to 145 km against smaller "boat size" targets. Minimum launch range is stated as 5.25 km. The Kh-59MK can be launched from the carrier aircraft when flying at speeds of Mach 0.5 to Mach 0.9 (600-1000 km/h) and from altitudes of 0.2 to 11 km, the missile flying to the target area at speeds of 900-1050 km/h at altitudes of 10-15 m over the sea, dropping to 4-7 m when in the terminal phase of the flight.

The Kh-59MK anti-ship missile, which is guided to the target by the ARGS-58E active radar homing head, is capable of engaging seaborne targets out to ranges of 285 km. NPO Saturn

Kh-29

The Kh-29TE(L) are short range air to surface missiles, six of which can be carried by the Su-30MK family; one on each of the intake stations, one on each of the inner wing stations and one on each of the intermediate wing stations.

These weapons are designed for use against hardened targets such as large bridges, reinforced runways, industrial centres and aircraft housed in hardened aircraft shelters, and can also be employed effectively against surface vessels with a displacement up to 10,000 tons. The Kh-29 missiles, which are 3.9 m in length, 0.4 m diameter, 1.1 m wing span and have a launch weight of 690 kg for the Kh-29TE and 660 kg for the Kh-29L, are carried on and launched from AKU-58AE airborne ejector units; the Kh-29TE being guided to the target by a passive TV guidance system whilst the Kh-29L is fitted with a semi-active laser guidance system; the target being destroyed by the 320 kg high explosive penetrating warhead. JSC Tactical Missile Corporation documentation shows the missile to have a minimum engagement range of 3 km and a maximum engagement range of 2-30 km for the Kh-29TE (depending on launch altitude) and 10 km for the Kh-29L.

A Kh-29 air to surface missile is launched from the third Su-30MKK development aircraft during weapons integration testing. Sukhoi

The Su-30MK family can also employ a number of guided and unguided bombs and rocket systems. Guided bombs include the KAB-500KR(OD) weapons, six of which can be carried on inner wing, intermediate wing and intake stations. A maximum of three of the larger KAB-1500KR(LG) weapons can be carried; one on the fuselage centre station and one on each of the inner wing stations.

Although not specified in released Sukhoi documentation, it is possible that the LGB-250 smart bomb will be integrated with Russian Federation Air Force Su-30M2 strike fighters, six or perhaps eight (this latter number is specified for the Su-35S) could be carried.

The Su-30MK family can carry eight FAB-500 (ZB-500, RBK-500, BetAB-500) general purpose bombs; two on each of the intermediate wing stations, two on fuselage centre station and one on each of the intake stations. Twenty eight of the smaller FAB-250 (OFAB-250-270) general purpose bombs can be carried in clusters on the same stations as used for the FAB-500 series. Alternatively the same stations can be used to carry up to thirty two OFAB-100-120 general purpose bombs.

Unguided rockets can include up to four B-8M-1 rocket pods; two on each of the intermediate wing stations or four smaller B-13L rocket pods can be carried on the same stations. Another alternative is four S-25, S-25OFM-PU rockets which are carried on the same stations as the B-8M-1 and B-13L. KnAAPO documentation shows that up to eight of the P-50T weapons can be carried on the intermediate wing stations (two each), fuselage centre station and engine intake stations.

JSC Tactical Missiles Corporation KAB-500 (left) and KAB-1500 (right) guided bombs are the standard medium and heavy guided bomb family on all members of the Su-30MKK/MK2/M2 family. *Author*

	KAB-500Kr	KAB-500-OD	KAB-1500Kr
Launch weight:	520 kg	370 kg	1525 kg
Warhead weight:	380 kg	250 kg	1170 kg
High explosive:	100 kg	250 kg	440 kg
Length:	3.05 m	3.05 m	4.63 m
Diameter:	0.35 m	0.35 m	0.58 m
Empennage:	0.75 m	0.75 m	0.85 m (folded)
Release altitude:	0.5-5 km	0.5-5 km	1-8 km
Carrier speed:	550-1100 km/h	550-1100 km/h	550-1100 km/h
Root mean Square deviation:	4…7 m	4…7 m	4…7 m
Warhead type:	concrete piercing (high explosive penetrator)	high explosive fuel air	high explosive

	KAB-1500LG-PrE	KAB-1500LG-F-E	KAB-1500LG-OD-E
Launch weight:	1525 kg	1525 kg	1450 kg
Warhead weight:	1120 kg	1120 kg	1170 kg
High explosive:	210 kg	440 kg	650 kg
Length:	4.28 m	4.28 m	4.24 m
Diameter:	0.58 m	0.58 m	0.58 m
Wingspan:	0.85 m (retracted) 1.3 m (extended)	0.85 m (retracted) 1.3 m (extended)	0.85m (retracted) 1.3 m (extended)
Release altitude:	1-8 km	1-8 km	1-10 km
Aircraft drop speed:	550-1100 km/h	550-1100 km/h	550-1100 km/h
Aiming accuracy:	4-7 m	4-7 m	4-7 m
Warhead:	penetrator	high explosive	fuel air explosive
Fusing:	contact with three types of delay	contact with three types of delay	direct action contact

Top: The R-27ER1 is in widespread Russian Federation Air Force service on various Sukhoi and MiG combat aircraft. The weapon is also operated by export Su-30 operators. Above: The R-27ET1 (left) infrared guided air to air missile allows the Su-30, as well as other Russian aircraft, an extended range passive (infrared) air to air capability compared to their Western counterparts. The only comparable missile in the west is the Mica IR arming the Dassault Rafale and Mirage 2000-9. Author

In the air to air role the Su-30MK2/M2 can be armed with the standard Russian medium and short-range air to air weapons; the Vympel (JSC Tactical Missiles Corporation) R-27ER1(R1), R-27ET1(T1), R-27EP1(P1), RVV-AE and R-73E (Note: Various in-house documentation refer to these weapons under differing but similar designations; for instance ER or RE. In the case of the ET1 and EP1 these are also sometimes referred to as ET/EP minus the 1. In Russian language the weapons are P-27, P-73 or K-27, K-73). In the case of the Su-30MKK it appears that the R-27R1 is the standard medium range air to air weapon. Certainly with the Su-30M2 variant, in service with the Russian Federation Air Force, the full spectrum of R-27 variants can be employed.

Entering service in the mid-1980's as the primary air to air armament of the Su-27S, the R-27 medium-range missile variants in service in 2015 are more capable updates of the R-27, of which a whole family of variants was produced, including the R-27R, NATO reporting name AA-10 'Alamo' A with SARH (Semi-Active Radar Homing) guidance and the R-27T 'Alamo' B with IR guidance. Longer range variants were also developed, designated R-27ER1 for the SARH variant and R-27ET1 for the infrared guided variant. These missiles, 'Alamo' C and 'Alamo' D respectively, are fitted with a boost sustain motor to extend engagement range.

Up to six R-27ER1(R1) missiles can be carried by the Su-30MK2; two on the fuselage centre stations, one on each of the engine intake stations and one on each of the intermediate wing stations. A maximum of two R-27ET1(T1) or R-27EP1(P1) missiles can be carried instead of the R-27ER1(R1); one on each of the intermediate wing stations. The R-27EP1(P1) is an air to air anti-radar missile designed to home in on opposing aircraft radar emitting in the forward hemisphere, a distinct advantage Russian fighter aircraft possess over their NATO opposite numbers.

Complementing the larger infrared guided R-27ET1 is the smaller, shorter range, but highly agile, Vympel (JSC Tactical Missiles Corporation) R-73E (NATO reporting name AA-11 'Archer') infrared guided missile, six of which can be carried by the Su-30MKK/MK2 family; one on each of the intermediate wing, one on each of the outer wing and one on each of the wingtip stations.

When it entered service in the 1980's, the R-73 was probably the most advanced short-range air to air missile in the world, being a generation ahead of the latest variants of the American AIM-9L/M Sidewinder or European Matra Magic 2 short-range infrared guided air to air missiles then arming NATO fighters. Only in the early 21st century did NATO field comparable systems in the shape of the MBDA (Matra British Aerospace Dynamics Alenia) ASRAAM (Advanced Short Range Air to Air Missile) and Raytheon AIM-9X Evolved Sidewinder.

The R-73 was developed with high agility as a design driver, augmented by the ability of the pilot of the Su-27 or MiG-29 fighters to cue the weapon to targets at up to 60° off-boresight via a HMSS (Helmet Mounted Sight System). High manoeuvrability was achieved by a combination of a number of factors, including four forward control fins, elevators attached to the rear fins, which are fixed, and deflector vanes positioned in the nozzle of the rocket engine.

The R-73E has a longer reach than most western equivalents such as the many AIM-9 variants; confirmed minimum engagement range against a tail-on target being 0.3 km with a maximum range of 30 km against a head-on target, with the capability of engaging targets manoeuvring at up to 12 g.

Model of a Su-30MKK configured with a pair of R-73E short-range infrared guided air to air missiles on the wingtip stations and another pair on the intermediate wing stations. Author

RVV-AE

Up to six RVV-AE active radar guided medium range air to air missiles can be carried by the Su-30MK2 family on the same stations as those used for the carriage of the R-27ER1(R1). Development of this weapon apparently commenced in 1982 and the missile began entering limited service in the mid-1990's, certainly with trials units. Into the 21st century the weapon has been integrated onto a number of aircraft types undergoing updates as well as new aircraft of the Su-27, Su-30, Su-34, Su-35S, MiG-29, MiG-35 variants as well as the Sukhoi T-50 PAK FA fifth generation multi-role fighter aircraft. The weapon has also been exported to a number of customers, including India, Malaysia, Peru, and China also apparently purchased the missile for its Sukhoi Su-30MKK/MK2 multi-role fighters.

The RVV-AE has narrow-span wings of rectangular shape and four lattice control surfaces at the rear; among the benefits of this type of control surface being reduced flow-separation at high angle of attack.

Like the US Raytheon AIM-120 AMRAAM (Advanced Medium Range Air to Air Missile) and the European MBDA MICA EM active radar guided missiles, the RVV-AE can be employed in a launch-and-forget mode and features a multi-stage guidance system that includes inertial in the initial phase with mid-course updates via an aircraft to missile datalink for long-range engagements, with active radar homing in the terminal phase of the engagement. The missiles on-board active-radar apparently has an acquisition-range of around 20 km.

Vympel noted that while the RVV-AE is heavier than the AIM-120A/C and MICA EM, the Russian missile has a longer range and better performance when engaging manoeuvring targets compared to its western rivals. The standard RVV-AE has a minimum engagement range of 0.3 km in the rear hemisphere and a maximum range of 80 km in the forward hemisphere, apparently reaches speeds of Mach 4 and can engage targets manoeuvring at up to 12 g from 0.2 to 25 km altitude. The missile, which features an active-radar fuse for the 22.5 kg warhead, can also apparently be used in a 'self-defence' mode to intercept missiles launched at the mother aircraft.

The R-73E is the standard short-range air to air missile arming Russian 4th, 4th+ and 4th++ generation combat aircraft. This example is shown on a wingtip station of a MiG-AT configured as a light combat aircraft. Author

Although not currently specified in released documentation, it is expected that the RVV-MD and RVV-SD, respective replacements for the R-73E and RVV-AE will be integrated with Su-30 variants.

An evolution of the R-73, the RVV-MD is a new generation highly agile infrared guided missile developed to arm the new generation of Russian 4th++ and 5th generation fighter aircraft. The JSC Tactical Missiles Corporation description states the "short range missile for close high manoeuvrable air combat provides hitting air targets (fighters, bombers, combat aircrafts, military aircrafts and helicopters), day and night, at all angles, on background of earth, under active enemy counteraction." The missile, which is powered by a single mode engine, features enhanced anti-jamming protection over its forebear, including optical jamming, and features "all angles passive infrared target homing (double range individual homing) with combined aero-gas dynamics control." The target is destroyed by a rod-shaped warhead activated by a laser non-contact sensor fuse in the RVV-MDL variant or a radio non-contact sensor in the RVV-MD. On Sukhoi and MiG fighter aircraft the weapon is carried on and launched from P-72-1D (P-72-1BD2) type rail tracked launcher.

A Vympel (JSC Tactical Missiles Corporation) RVV-AE active radar guided air to air missile is launched from a Su-27UB trials aircraft.

The RVV-SD, developed by Vympel (JSC Tactical Missiles Corporation), is clearly an evolution of the RVV-AE incorporating a number of improvements over its forebear, with longer engagement range, increased engagement capability and enhanced resistance to electronic countermeasures. Tactical Missiles Corporation describes the missile as "intended for hitting air targets (fighters, bombers, attack aircraft, helicopters… cruise missiles) day and night, at all angles, under electronic countermeasures, on background of earth and water surfaces, including multichannel application 'fire-and-forget'". The missile, which is powered by a single mode rocket engine, incorporates inertial homing "with radar correction and active radar self-homing". The target is destroyed by a rod-shaped multi-charge warhead with detonation by laser non-contact target sensor. For external carriage on 4th, 4th+, 4th++ and 5th generation aircraft the missile is carried on and launched from the AKU-170E missile ejection launcher.

Vympel R-73E (top) and RVV-AE (centre) bestow upon the carrier aircraft a highly effective medium and short-range air to air capability. These weapons will be superseded by the RVV-MD and RVV-SD respectively. Author

Vympel R-27

Propulsion: two mode solid propellant rocket motor
Length: 4.775 m for R-27ER1 and 4.49 m for R-27ET
Diameter: R-27ER1 and ET 0.26 m at solid rocket section and 0.23 m at control unit section
Span: wing 0.803 m and control plane 0.972 m
Launch weight: R-27ER1 350 kg, R-27ET 343 kg
Speed: Mach 4
Range: R-27ER1 – 60-62.5 km against fighter aircraft sized targets and up to 100 km against larger targets; R-27ET – 80 km in front hemisphere
Warhead: 39 kg expanding rod
Guidance: R-27R and R-27ER1 (SARH), R-27T and R-27ET (passive infrared)

Vympel R-73E

Propulsion: solid propellant rocket motor
Length: 2.9 m
Diameter: 0.17 m
Span: 0.51 m fin span and 0.38 m control plane span
Launch weight: 105 kg
Range: 30 km maximum head on and 0.3 km minimum tail on against up to 12 g manoeuvring targets
Engagement altitude: from 0.02 to 20 km
Warhead: 8 kg high explosive expanding rod
Guidance: all-aspect passive infrared

Vympel RVV-AE

Propulsion: solid propellant rocket motor
Length: 3.6 m
Diameter: 0.2 m
Wingspan: 0.4 m
Control plane span: 0.7 m in flight position
Launch weight: 175 kg
Speed: Mach 4 class
Range: minimum 0.3 km in rear hemisphere and maximum 80 km in front hemisphere
Engagement altitude: 0.2 to 25 km
Warhead: 22.5 kg high explosive
Fuse: active-radar
Guidance: inertial, command and active-radar in the terminal phase

6

Su-35UB – SUKHOI ENIGMA

The Sukhoi Su-35UB, which could certainly be regarded as something of an enigma, is included in this volume as the aircraft is a two-seat multi-role fighter in the same mould as the Su-30MKK/MK2 series. The UB designation would indicate that the aircraft had an operational conversion trainer role for any planned production batches of Su-27M (original Su-35) canard-tri-plane single-seat multirole fighters. However, the aircraft features an advanced modern avionics and fire control system allowing it to conduct the full spectrum of missions of a multi-role strike fighter. Unlike previous variants of the 'Flanker' family, KnAAPO utilised large-scale use of digital design tools for the Su-35UB.

Resembling a Su-30MK2 variant with the taller square topped vertical tail fins, carried over from the Su-27M, but apparently slightly modified, the Su-35UB can easily be distinguished from the Su-30MKK by the inclusion of canard foreplanes. The aircraft was initially powered by two standard AL-31F engines.

As with the Su-30MKI/SM variants manufactured at Irkut, the Su-35UB was apparently equipped with the N011M Bars radar system allowing the use of the full range of ordnance associated with the Su-30MKI/MKK series, which can be carried on the twelve external stores stations. The aircraft features an in-flight refuelling probe on the port side forward fuselage necessitating the OLS to be offset to starboard in front of the windscreen as on the Su-30/33/35S.

The Su-35UB could also be described as an advanced evolution of the Su-30MK2 to rival the Irkut built Su-30MKI series, which, unlike the KnAAPO built Su-30MK2, were, as noted above, equipped with the modern Bars multi-mode radar system and canard foreplanes, the former increasing the aircraft capabilities in the air to air and air to surface role while the latter, combined with thrust vector control for the engine nozzles, enhanced aircraft handling and manoeuvrability. With the success of the Su-30MKI series the requirement for an aircraft in the class of the Su-35UB seemed redundant and only a single demonstrator, coded Blue 801, was built, this conducting its maiden flight in autumn 2000.

The single Su-35UB, which conducted its maiden flight in 2000, has been used by KnAAPO as a demonstration and trials aircraft. Sukhoi.

APPENDICES

APPENDIX I

Su-30MKK Development Aircraft	
T-10PU-5	Su-30 development aircraft, Blue 05, converted to act as a pseudo Su-30MKK systems trials aircraft
Blue 501	First Su-30MKK development aircraft
Blue 502	Second Su-30MKK development aircraft
Blue 503	Third Su-30MKK development aircraft
Blue 504	Fourth Su-30MKK development aircraft

APPENDIX II

Su-30MKK/MK2/M2 Operators			
	Su-30MKK	Su-30MK2	Su-30M2
China	76	24	
Indonesia		5	
Vietnam		24	
Venezuela		24	
Russian Federation			20

APPENDIX III

Su-30MK2 Specification (Sukhoi/UAC data)

Powerplant: two x AL-31F bypass turbofan engines each rated at 12500 kgf with afterburner
Length: 21.9 m
Height: 6.4 m
Wing span: 14.7 m
Normal Take-off weight: 24900 (figures for Su-30MK) configured with 2 x R-27R1, 2 x R-73 air to air missiles and 5720 kg of fuel
Maximum take-off weight: 34500 kg with a 38000 kg absolute limit
Landing weight: 23600 kg normal and 30000 kg maximum
Internal fuel load: 5720 kg normal and 9720 kg maximum (conflicting documentation states 9640 kg maximum)
Maximum level speed: 1400 km/h at sea level, 2100 km/h at high altitude for Su-30MK2 and 1350 km/h at sea level and 2120 km/h at latitude for Su-30MK(I) series
Maximum Mach number: 2
Ceiling: 17300 m without external stores
Range with maximum fuel load: when expending 2 x R-27R1 and 2 x R-73E air to air missiles "at half distance", 1270 km at sea level, 3000 km at altitude without in-flight refuelling and 5600 km with one in-flight refuelling (figures for the Su-30MK(I) are 3000 km and 5200 km respectively for the latter two values)
Maximum airborne time: Pilot dependant, but generally stated as 10 hours
In-flight refuelling system: 100 l/m at "maximum flow rate (at entry pressure of 3.5 kg/cm^2)
Take-off run: 550 m at normal take-off weight
Landing roll: 750 m with deployment of a braking parachute
Load limit: 9 g.
Crew: 2
Fixed armament: GSh-301 30 mm cannon housed in port wing root with 150 rounds of ammunition
Maximum external stores load: 8000 kg carried on twelve external stations. Can include various combinations of R-27ER1(R1), R-27ET1(T1), R-27EP1, RVV-AE, R-73E air to air missiles, Kh-29TE(L), Kh-31P(A), Kh-35E and Kh-59ME/MK air to surface missiles, KAB-500Kr(OD) and KAB-1500Kr(LG) guided bombs and various combinations of unguided bombs and rockets.
Service life: 3000 hours
Time to first overhaul: 1500 hours
Engine and outboard accessory gearbox life: 500 hours to first overhaul and 1500 hours service life

APPENDIX IV

Su-27S Specification (Sukhoi data)

Powerplant: 2 x AL-31F afterburning turbofan engines each rated at 12500 kgf -2% in afterburner and 7670 kgf in full military power +/-2%
Length: 21.9 m
Wingspan: 14.7 m
Height: 5.9 m
Normal take-off weight: 23400 kg with 2 x R-27R1 and 2 x R-73E missiles and 5270 kg of fuel
Maximum take-off weight: 30450 kg
Maximum landing weight: 23000 kg
Maximum internal fuel load: 9400 kg
Maximum speed at sea level: 1400 km in clean configuration
Maximum Mach number: 2.35
Ceiling: 18.5 km in clean configuration
Operational range configured with 2 x R-27R1 and 2 x R-73E air to air missiles when the missiles are launched at "half distance": 1340 km at sea level and 3530 km at altitude
Take-off run: 450 m at normal take-off weight
Landing run with braking parachute: 620 m at normal landing weight
Load limit: 9 g
Crew: 1

Su-27UB (Specification is the same as Su-27S except in the following)

Height: 6.4 m
Normal take-off weight: 23900 kg with 2 x R-27R1 and 2 x R-73E missiles and 5270 kg of fuel
Maximum take-off weight: 33000 kg
Maximum Mach number: 2.0
Ceiling: 17.5 km in clean configuration
Operational range configured with 2 x R-27R1 and 2 x R-73E air to air missiles when the missiles are launched at "half distance": 1270 km at sea level and 3000 km at altitude
Take-off run: 550 m at normal take-off weight
Landing run with braking parachute: 670 m at normal landing weight
Crew: 2

GLOSSARY

AIM	Airborne Interception Missile
AMRAAM	Advanced Medium Range Air to Air Missile
AoA	Angle of Attack
ASRAAM	Advanced Short-Range Air to Air Missile
AWACS	Airborne Warning and Control System
CIS	Commonwealth of Independent States
cm^2	Centimetre Squared
Cobra	Extreme high angle of attack manoeuvre
DBS	Doppler Beam Sharpening
ECM	Electronic Counter Measures
EDCS	Electronic Distance Control System
ELINT	Electronic Intelligence
F	Fighter
FADEC	Full Authority Digital Engine Control
FBW	Fly By Wire
FCS	Flight Control System
FX	Fighter Experimental
G	Gravity (1 G = 1 x Earth gravity)
g	Gravity (1 g = 1 x Earth gravity)
GLONASS	Globanaya Navigozionnaya Sputnikovaya Sistema (Global Navigation Satellite System)
GMTS	Ground Moving Target Selection
HMSS	Helmet Mounted Sight System
HMTDS	Helmet Mounted Target Designation System
HP	High Pressure
HRH	High Resolution mode
HUD	Heads Up Display
IA-PVO	*Istrebitelnaya Aviatsiya Protivo-Vozdushnoy Obstrany*/Air Defence Force
IAPA	Irkut Aircraft production Association
IFF	Identification Friend or Foe
IRHH	Infra-Red Homing Head
IRST	Infra-Red Search and Track
kgf	Kilogram Force
kg/s	Kilogram per second
km	Kilometre
km/h	Kilometre per Hour
kN	Kilo Newton
KnAAPO	Komsomolsk-on-Amur Aircraft Production Association
LERX	Leading Edge Root Extension
l/m	Litre per Minute
LP	Low Pressure

m	Metre
m²	Metre Squared
Mach	Speed of Sound
MBDA	Matra British Aerospace Dynamics Alenia
MFDC	Multi-Function Display Screen
MFLCDS	Multi-Function Liquid Crystal Display Screen
MiG	Mikoyan
MRCA	Multi-Role Combat Aircraft
MRH	Medium Resolution mode
NATO	North Atlantic Treaty Organisation
NAVSTAR	Navigation Satellite Timing and Ranging System
OEPS	Optical Electronic Sighting System
OLS	Optical Location System
PFI	Advanced Frontline Fighter
PLAAF	Peoples Liberation Army Air Force
PLNAF	Peoples Liberation Naval Air Force
RB	Real Beam mode
RCS	Radar Cross Section
RHH	Radar Homing Head
s	Second (unit of time)
SAR	Synthetic Aperture Radar
SARH	Semi-Active Radar Homing
SS	Sea Search
Su	Sukhoi
Tail Slide	Extreme high angle of attack manoeuvre
TBO	Time between Overhaul
TNI	Indonesian Military
TV	Television
TWS	Track While Scan
US	United States
USAF	United States Air Force
v	Velocity
WCS	Weapon Control System
x	Times (multiplication)

ABOUT THE AUTHOR

Hugh, a historian and author, has published in excess of forty books; non-fiction and fiction, writing under his given name as well as utilising two different pseudonyms. He has also written for several international magazines, whilst his work has been used as reference for many other projects ranging from the aviation industry, international news corporations and film media to encyclopaedias, museum exhibits and the computer gaming industry. He currently resides in his native Scotland

Other titles by the Author include

Sukhoi T-50/PAK FA - Russia's 5th Generation 'Stealth' Fighter
Sukhoi Su-35S 'Flanker' E - Russia's 4++ Generation Super-Manoeuvrability Fighter
Sukhoi Su-34 'Fullback'
Eurofighter Typhoon - Storm over Europe
Tornado F.2/F.3 Air Defence Variant
Boeing Super Hornet and Growler
Air to Air Missile Directory
Boeing X-36 Tailless Agility Flight Research Aircraft
X-32 - The Boeing Joint Strike Fighter
X-35 - Progenitor to the F-35 Lightning II
X-45 Uninhabited Combat Air Vehicle
Into The Cauldron - The Lancaster MK.I Daylight Raid on Augsburg
Light Battle Cruisers and the Second Battle of Heligoland Bight
British Battlecruisers of World War 1 - Operational Log, July 1914-June 1915
Hurricane IIB Combat Log - 151 Wing RAF, North Russia 1941
RAF Meteor Jet Fighters in World War II, an Operational Log
Typhoon IA/B Combat Log - Operation Jubilee, August 1942
Defiant MK.I Combat Log - Fighter Command, May-September 1940
Blenheim MK.IF Combat Log - Fighter Command Day Fighter Sweeps/Night Interceptions, September 1939 - June 1940
Tomahawk I/II Combat Log - European Theatre - 1941-42
Fortress MK.I Combat Log - Bomber Command High Altitude Bombing Operations, July-September 1941
North American F-108 Rapier
F-84 Thunderjet - Republic Thunder
USAF Jet Powered Fighters - XP-59-XF-85
XF-92 - Convairs Arrow

www.ingramcontent.com/pod-product-compliance
Lightning Source LLC
Chambersburg PA
CBHW041523220426
43669CB00003B/36